INNOVATION RIGHTS

... and Wrongs

Donald F. Moyer, PhD

Specialty Publications
Oregon Digital Design, LLC
publish@oregondigitaldesign.com

Printed in the U.S.A

Contents

Preface

Dramatic, learn-by-doing, simplified mock trial competitions are powerful tools for achieving several primary goals:

Helping the greatest diversity of persons understand innovation rights is the best way to encourage innovation.

Participants learn to act effectively where slippery issues must be decided and the "facts" can be interpreted in contradictory ways.

Though legal training is not a goal, participants do learn courtroom etiquette and professional and success-positive behaviors.

No legal training at all is needed to use these scenarios.

These mock trials can be used by all kinds of persons in all sorts of environments.

Innovation rights include more than patent rights. "Right" has many and various meanings such as "just and good" such as "correct" such as "genuine, real" such as "preferable" such as "best practices" and more. These various other meanings of "right" are in play by the mock trial scenarios.

Since, for every trial there is necessarily at least one innovation wrong which led to the trial, many kinds of innovation wrongs are illustrated by the trials and the backgrounds for the trials.

In the original version of these mock trials the attorney and witness players and the jury were high school students at Austin Polytechnical Academy in Chicago's Austin neighborhood. The experienced persons being coaches, judge, and bailiff were law students from Chicago's John Marshall Law School. See Appendix E for acknowledgements.

Thanks to the work of those high school students and those law students, we ironed out many wrinkles to create a very smooth running model for these mock trials.

Now this mock trial model and these mock trial scenarios can be used by many others in many other settings.

Persons could use this model on their own. The model could be used by teachers and other formal and informal mentors.

Younger persons could be players in this model. Older persons could be players in this model.

The model and scenarios can work in various learning environments such as schools, museums, libraries, churches, peer-to-peer mentoring.

I can't imagine all the environments where this model and these scenarios will work, nor can I imagine and all the age groups which can enjoy the model and scenarios.

www.InnovationRights.org

General Introduction

Trying to explain the importance of innovation rights, often some smarty-pants would say with great certainty that a patent is just a right to sue. The response is: Yes, and you're likely to be the one to be sued.

If you don't understand innovation rights – and don't understand the many wrong things you can do to lose rights – then your bright ideas likely will be wasted.

These mock trials show many ways the gavel can come down hard on smarty-pants, many ways smarty-pants can lose valuable innovation rights.

These mock trial scenarios include arguments about whether an invention is, or is not: infringed, novel, obvious, enabled, invalid because of the on-sale bar, invalid because of inequitable conduct, the property of an employer, and include arbitrations related to license agreements and joint venture agreements.

Innovation rights include more than patent rights. "Right" has many and various meanings such as "just and good" such as "correct" such as "genuine, real" such as "preferable" such as "best practices" and more. These various other meanings of "right" are also in play in the mock trial scenarios.

Since, for every trial there is necessarily at least one innovation wrong which led to the trial, many kinds of innovation wrongs are illustrated by the trials and the backgrounds for the trials.

To understand the context in which the mock trials take place, here, in this general introduction, we will:

1) Clarify the terms we'll use in the mock trials, such as what a patent is and what enforcement of a patent is.

2) Explain the structure of trials and the roles of advocate and witness.

3) Explain the structure of the scenarios.

Then, in the introduction to Part One, we will meet the main actors in our scenarios. Basic innovation rights issues are the focus of Part One. These will be summarized in the Part One introduction.

In Part Two the main actors from Part One interact along several story lines to show how various decisions by these actors can cause lasting innovation wrongs. The Part Two introduction is a synopsis of these story lines and the ensuing mock trials.

In these mock trials typically a plaintiff has innovation rights – a patent for example – and sues a defendant to enforce that right. A plaintiff typically will argue in a trial that the plaintiff has a valid right and that the defendant is infringing that right.

Defendant might argue in the trial that the defendant is not infringing. Alternatively the defendant might argue in the trial that the right asserted by the plaintiff – typically a patent – is not valid because the plaintiff failed to meet all the requirements for the right.

When the defendant argues that the right is invalid, the plaintiff will counter that argument.

First let's be clear about meanings.

We use the economic sense that innovations are new products and processes which create new value.

Inventions are new products and processes claimed in valid patents.

Very few inventions create new value so most inventions are not innovations in the economic sense.

And, innovations need not be inventions. You can think of some important innovations which are not claimed in a valid patent. The Internet for example.

Though not all innovations are inventions, it is easy to see how patents are very important for the success of innovations:

1) You have an idea for a new product.

2) I have money to invest.

3) Why should I invest in your idea?

4) I want to know how we could stop others from copying your idea.

5) I want to know if you have preserved your patent rights.

6) I want to know if your idea is ready for patenting.

7) If you can give good answers to my questions I might invest.

8) Why do I think that patent rights are so important for innovation?

9) If your idea for a new product does create new value, such as profits for you and me, then copycats will want some of the loot.

10) With a patent we can stop copycats from taking any of our well-deserved profits.

The answer to how we might be able to stop copycats from taking any of our profits starts with the **US Constitution -- Article 1, Section 8, Clause 8:**

"Congress shall have the power to . . . promote the Progress of Science and useful Arts, by securing for limited Times to Authors and Inventors the *exclusive Right* to their respective Writings and Discoveries" *[emphasis added]*

Using this power, Congress passed the law – Title 35 of the US Code – establishing the US Patent and Trademark Office to issue patents which meet all requirements of the law.

On the face of a patent are these words: "this United States Patent grants to the person(s) having title to this patent the right to *exclude others* from making, using, offering for sale, or selling the invention throughout the United States or importing the invention into the United States, and, if the invention is a process, of the right to exclude others from using, offering for sale or selling throughout the United States, or importing into the the United States, products made by that process . . ." *[emphasis added]*

A patent is *infringed* when someone does any of the underlined things in the paragraph just above. When a patent is infringed, then, as the law states:

"A patentee shall have remedy by civil action for infringement of his patent."

To enforce the right by civil action the patent owner – the plaintiff – sues the alleged infringer – defendant – in a federal court. These mock trials are simplified examples of this civil action.

An invention is what is claimed in a valid patent. Let's look at the example in Appendix A to see what that means. The patent copied there is that of an inventor who made the invention to ease his work unloading tires from trucks onto the conveyor at the venerable Ford automobile manufacturing plant in south Chicago.

Think someone doing that kind of work is unlikely to make an invention? Wrong!

Vernon's new result is an unexpected discovery not anticipated by any prior art. (In patent lingo "prior art" is anything which anticipates the idea because it was already known when a patent application is filed. There is more about finding prior art in Appendix C.)

In Appendix A you see first the face of the published patent. The stuff inside the document shown in the photo on the cover of this book, with the gold seal and red ribbon, is the published patent.

The face of the published patent is full of very useful information which is described more fully in Appendix B.

Just after the face are the drawings. Only one sheet of drawings here. Sometimes there are many sheets of drawings.

After the drawings is the written description. In this case it is just one page long – usually much longer. The first paragraph tells what the invention is. The second paragraph notes that the invention is a new result, and unexpected discovery.

In this patent the last page of the patent has all the claims. The first claim is the broadest. Then there are two claims which depend from the first and introduce new things. The fourth claim includes everything.

Claims two and three are "dependent claims" meaning that they depend from claim 1, meaning that the dependent claims have all the stuff of the claim from which they depend plus new stuff. Thus, claim two has all the stuff of claim one plus new stuff, and claim three has all the stuff of claim one and claim two plus new stuff.

If a competing product has all the stuff, and maybe more, in <u>any</u> of Vernon's claims, then that product infringes Vernon's patent. Vernon could sue the maker of this infringing product to stop them from infringing and to get damages.

Which is what usually happens in these mock trial scenarios:

1) A plaintiff has a patent

2) The plaintiff sues a defendant who is selling a competing product.

3) The defendant might argue that the defendant is not infringing, which happens in the first scenario below.

4) The defendant might argue that the plaintiff's patent is not valid for reasons argued in other scenarios.

In these mock trials there are several roles which to be taken by players:

1) The attorney for the plaintiff has four roles:

a) Opening statement.
b) Direct examination of the plaintiff's witness.
c) Cross examination of the defendant's witness.
d) Closing argument.

One player can take all these roles, each of the four roles can be taken by each of four players, and the four roles can be distributed among two or three players.

2) Another role, taken by another player, is the witness for the plaintiff. Two players could share the role as witness, one doing the direct and another doing the cross examination.

3) The attorney for the defendant has four roles:

a) Opening statement.
b) Direct examination of the defendant's witness.
c) Cross examination of the plaintiff's witness.
d) Closing argument.

One player can take all these roles, each of the four roles can be taken by each of four players, and the four roles can be distributed among two or three players.

4) Another role is, taken by one player, is witness for the defendant.

5) Another player can be a judge.

6) Another player can be bailiff keeping order.

7) Other players can be jurors.

8) Other players, typically experienced persons such as law students, can coach the attorney and witness players.

Each mock trial is itself a competition, which is part of the reason this is a powerful learning tool. Bigger competitions can be staged, with several rounds of trials, with teams switching sides (The plaintiff becomes the defendant and visa versa), with players on teams switching roles (attorneys become witnesses and visa versa). Many possibilities here all adding depth and breadth to the learning.

These simplified mock trials – which can take as little as half an hour – have the following parts:

1) The judge explains the trial to the jury.

2) The plaintiff's attorney makes an opening statement.

3) The defendant's attorney makes an opening statement.

4) The plaintiff's attorney takes direct testimony from the plaintiff's witness.

5) The defendant's attorney cross examines the plaintiff's witness.

6) The defendant's attorney takes direct testimony from the defense witness.

7) The plaintiff's attorney cross examines the defense witness.

8) A short recess for preparing closing arguments (three to five minute recesses work).

9) The plaintiff's attorney makes a closing argument.

10) The defendant's attorney makes a closing argument.

11) The judge instructs the jury.

12) The jury delivers a verdict.

13) Experienced persons can give feedback to the attorney and witness players.

The plaintiff and the defense teams, and any coaches, can be designated a week ahead.

At the conclusion of mock trials experienced persons can give the attorney and witness players feedback about:

1) How well each player understood the trial issues (it works well for players to have trial scenarios ahead of time so that players can understand the scenario rather than just quote the scenario).

2) How persuasive each player was in the trial (note how this is an acting exercise to appear to be more trustworthy than the other side).

3) How well each player responded to unexpected events (part of the game is to present the other side with something unexpected to upset their plan).

4) How well each player used success-positive behavior

5) How well each player observed court room etiquette.

These mock trial scenarios are the raw material which the several players use to prepare for their rolls in the mock trials. The scenarios comprise:

1) The story, telling how the trial came to be.

2) The part of law at issue simply stated, which the judge will explain to the jury.

3) What the jury must decide, which the judge will explain in the instructions to the jury.

4) The plaintiff's attorney's goals to be achieved in the opening statement, the direct examination of plaintiff's witness, the cross examination of the defendant's witness, and the closing argument.

5) The defendant's attorney's goals to be achieved in the opening statement, direct examination of the defendant's witness, cross examination of the plaintiff's witness, and closing statement.

6) Facts not in dispute.

7) Depositions (also facts not in dispute):

a) From the plaintiff's witness, questions asked by the plaintiff's attorney (background for direct testimony).

b) From the plaintiff's witness, questions asked by the defendant's attorney (background for cross examination).
c) From the defendant's witness, questions asked by the defendant's attorney (background for direct testimony).
d) From the defendant's witness, questions asked by the plaintiff's attorney (background for cross examination).

The statement of the part of law at issue is simplified to keep focus on one issue at a time. Real trials likely deal with messy combinations of issues.

The statement of what the jury must decide is also simplified to keep focus on the one issue. Innovation rights disputes can be very hard for juries because there are no smoking guns. The "facts" can be understood many ways.

The word "facts" just above was written with the scare quotes to make it clear that there is always some uncertainty, always some alternative explanation. The word "facts" is used here to indicate things which are in evidence and are not in dispute. Attorneys and witnesses can not change these "facts" and can not invent new "facts" beyond what is in evidence.

The goals for the plaintiff and the defendant teams are guides which the players will use in statements and testimony. Persons in the roles of attorney, should have a plan they are comfortable with. If not they may not be able to persuade the jury to find for their side.

The opening statements outline a picture showing an interpretation of the "facts" which will lead – they hope – to a favorable decision by the jury. Testimony will provide evidence for the picture. So, an attorney and the attorney's witness should work as a team to plan this picture and to paint the picture.

The witness on an attorney and witness team should also be comfortable with the plan. The attorneys and witnesses should not read from the depositions. They should be able to ask and answer in their own words. Otherwise they will not be persuasive. <u>It is about acting</u>. It is about <u>appearing to be</u> more trustworthy than the other side.

In cross examinations attorneys will try to make witnesses look untrustworthy so for cross examinations witnesses should be ready for unexpected questions. Witnesses should pay close attention to direct testimony by the other side witness in order help their attorney come up with unexpected questions for cross examination of the other side witness. It is about team work.

If a witness gives an answer which only seems to be different from what is in the "facts" of the case then, the witness could look untrustworthy when challenged about the difference. Pointing out differences is one way to upset the picture being painted by the other side.

In the direct testimony witnesses are the stars. The attorneys are just trying to make the witnesses look good.

In cross examination attorneys are the stars trying to make the witnesses look bad. In cross examinations attorneys should ask questions which are not expected by witnesses. The witness will look good by responding well. The witness will look bad if the witness is confused.

Another way to upset the other side is with objections to questions by the other side attorney and with objections to answers by the other side witness.

For example, attorneys can object to questions which:

1) Lead the witness in direct testimony.

2) Ask for speculation.

3) Are argumentative rather than asking for information.

For example, attorneys can object to answers which:

1) Go beyond the witnesses direct knowledge.

2) State conclusions rather than evidence.

3) Are not about "facts" of the case (such as calling someone a jerk).

Witnesses can prompt their attorney counterpart to object.

After the testimony part of the trial there will be a short recess for the two attorney and witness teams to prepare closing arguments. These closing arguments should:

1) Restate the picture painted in the opening statement and point out how the direct testimony provided evidence for the picture and show how the picture withstood all challenges by the opposing team.

2) Point out weaknesses in the picture painted by the opposing team, pointing out weakness in testimony by opposing witnesses and pointing out reasons the opposing witness should not be believed.

3) Ask the jury to find in favor of your side.

Courtroom Etiquette

A courtroom is a place reserved for discipline, respect, and orderly conduct. It is a part of our justice system, our constitution, and integral in the founding of our country. There are specific rules that apply when one goes to court and presents his or herself to a judge and a jury.

Patent law is a federal issue that is heard in federal court. In my personal experience, federal courts are generally more formal than state courts, but this depends heavily on the judge that is assigned to your case. The judge sets the tone of the courtroom. Some judges are more formal and may adopt firmer rules than others. It is best to be as formal as possible. If the judge you are before, calls for a more relaxed environment, then, and only then, would it be permissible to relax the rules.

Because these scenarios are only "mock" trials and are not before a real judge and a real jury, every single rule below might not apply. However, practice as if you are before a real judge who is sitting on a real bench in a real courtroom. For example:

Stand up straight and make eye contact.

Do not chew gum.

Be polished with your choice of clothing and your choice of words.

Dress in conservative business attire.

Utilize the English language and do not speak slang.

Address the judge as "Your Honor."

Speak only when the judge instructs you to or gives you permission.

Introduce yourself and the party you represent.

Speak confidentially.

Never interrupt the judge. If the judge interrupts you, do not speak over the judge.

Usually, there is a court reporter or a recording device in the courtroom, so it is best for only one person to speak at a time.

<u>Opening Statement Etiquette:</u>

Your honor, opposing counsel, ladies and gentlemen of the jury my name is ______________ and I represent ____________________. Today you will hear <u>[my wittness]</u> tell you that: Tell the jury exactly what they will hear. Then:

We ask that you find in favor of my client _________.

Direct Examination Etiquette:

The witness is the one telling the story. There should be a direct line of communication and eye contact from the witness to the jury box. Stay at the lectern – there is no reason to approach a witness in these mock trials.

Use your own words in order to frame this testimony to support what you want the jury to decide.

Cross Examination Etiquette:

This is your time to discredit the witness and make the witness appear dishonest and untrustworthy.

Closing Argument Etiquette:

Your honor, opposing counsel, ladies and gentlemen of the jury my name is ____________ and I represent ___________________. Retell the story your witness told, point out weakness of the other side's story and say:

We ask that you find in favor of my client _________.

Part One

Part One Introduction

Though innovation rights are more than patent rights, patent law is a key part of innovation rights.

In Part One we focus on practical patent law issues and focus on innovation wrongs which can lead to loss of patent rights, for example:

1) What a patent is and how a patent is enforced in an infringement trial.

2) That the claims in a patent must be novel.

3) That the claims in a patent must not be obvious changes in existing art.

4) That the patent must describe the invention to enable persons of ordinary skill to make and use the invention.

5) That the patent must disclose the best mode the inventor contemplates to make and use the invention.

6) That the invention must not be on sale more than one year prior to applying for a patent.

7) That employer's rights can mean that employers can own inventions made by employees.

8) That there must be no inequitable conduct before the USPTO.

First, let's introduce the players.

Quinn is a young person with a deep desire to start a business. Quinn has a good sense of what products could succeed. Also, Quinn has lots of ideas but does not have the skills to turn the ideas into products.

Quinn hired Brook to turn ideas into inventions. Remember that an invention is a new product or process claimed in a valid patent.

Brook is a successful inventor but is not good at the business part of innovation. Remember that an innovation is a new product or process which creates new value.

Because Quinn's business has no income, Brook is paid in stock. This is OK with Brook because together Quinn's skills and Brook's skills could make a profitable business. They named the company QUBCO pronounced cube-co.

Quinn's friend Shelby is forever losing things. This gave Quinn the idea for the first product called SmartCoat.

Shelby is a wild card in our scenarios.

Quinn's raw idea was to have SmartCoat give an alarm when the person wearing SmartCoat was leaving something behind. But, Quinn did not know how to make this work.

Brook invented the SmartCoat alarm system using RFID tags on items which might be lost and uses a base module in SmartCoat, to tell when an RFID tag is not close to SmartCoat.

Brook is patent savvy and convinced Quinn to apply for a patent on SmartCoat. In most of the scenarios in Part One a patent does issue to QUBCO.

Soon after Quinn and Brook began working together Shelby broke the connection with Quinn and went to work at JUGCO, a third generation clothing manufacturer.

JUGCO is pronounced jug - co. Emerson, the second generation head of JUGCO played jug in a jazz band.

When times got tough for clothing manufacturing, Emerson established a powerful innovation process at JUGCO in order to stay competitive.

JUGCO has a very diverse workforce which means that the workforce has many different ways to think about new products. Emerson made sure that all those very diverse workers have a strong sense of belonging to a JUGCO community and cooperate and work very well together.

Emerson introduced an element of competition so that JUGCO workers are rewarded for innovative ideas.

Emerson also made sure that respect for diversity and respect for competition which contributes to the common good are both strong JUGCO community values. This well balanced linkage of community, competition, cooperation, and diversity is powerfully innovative.

Rowan, head of JUGCO after Emerson passed away, continued Emerson's innovative policies. So, when Rowan learned about SmartCoat, JUGCO innovation went full speed ahead. Soon, JUGCO did have a better and less expensive product on sale, called SmartyCoat.

Our mock trial scenarios use several ways Quinn and JUGCO can confront each other to illustrate practical issues in patent law.

The infringe scenario is for a basic infringement trial where QUBCO sues JUGCO and JUGCO argues that they are not infringing. This scenario is a practical example of the part of the General Introduction above which explained the constitutional basis of patents and the enforcement of the patent right, which you may want to review.

The novelty and trade secret scenario is a practical example of the basic requirement for a patent. The constitutional bargain is that you promote progress by teaching us all something new and you get in return the right to exclude. The patent right can be lost even if someone steals trade secrets and discloses them.

The constitutional bargain is about discoveries. The obvious scenario is a practical example of the rule that improvements which are obvious to try and expected to work are not discoveries. So, these obvious improvements are not patentable.

The constitutional bargain requires that in return for a patent right you must teach us something new. The patent law thus requires that you describe your claimed invention sufficiently so that a person with appropriate skills can make and use the invention. The enable scenario is a practical example of this requirement.

The patent law also requires that you tell everyone in the patent specification the best way you intend to make and use the claimed invention. Under the new law passed late September 2011, it is not clear what might happen if you do not meet this best mode requirement. The license arbitration scenario grapples with this ambiguity.

A patent right – as specified in our Constitution – is for a limited time only. It would not be fair for you to make and sell a product for an unlimited time and only patent it when you wanted to exclude others. Thus the law has an on sale bar, meaning that, if your claimed invention is on sale for as long as a year before you apply for a patent, then you are bared from having a patent right. Our on sale scenario is a practical example of this.

The patent law is not the only law about innovation rights. State employment laws can specify that employees can be obliged to assign rights to inventions to their employer. Our employer rights trial is a practical example of this.

A patent right to exclude others from making and selling copies of your claimed invention is super valuable. That right can be lost many ways – there are more innovation wrongs than innovation rights. Our scenarios are practical examples of some innovation wrongs where that right can be lost.

Another way to lose a patent right is to engage in inequitable conduct with the USPTO. Withholding information relevant to any of the issues in the scenarios here could be inequitable conduct. The inequitable conduct scenario is a practical example of this and an example of how this can be very slippery.

SmartCoat Infringe Trial

Quinn wants to build a new business around smart products. The first smart product was inspired by Quinn's friend Shelby who forever forgets things such as gloves. The product was called SmartCoat, because SmartCoat gives an alarm when an item which should be near the coat – a glove, for example – is not near the coat.

Brook, an experienced and gifted inventor, joined Quinn to invent the details of SmartCoat and to invent other smart products. Since Quinn could only pay in stock they were co-owners of the business they called QUBCO.

Shelby went to work at JUGCO and soon thereafter JUGCO began work on SmartyCoat.

QUBCO began making and selling SmartCoat and soon thereafter JUGCO began making and selling SmartyCoat.

QUBCO had to delay filing a patent application until income from SmartCoat could cover the cost.

When QUBCO's SmartCoat patent was granted, QUBCO sued JUGCO for infringing the SmartCoat patent. (Remember the discussion in the General Introduction above about what a patent is, what infringement is, and how a patent is enforced.)

Your patent is infringed when someone sells a product which has all the parts in a claim in your patent. You must sue the infringer to make them stop and to get damages.

The key parts of the SmartCoat patent claims are an RFID tag secured at an item such as a glove, a base module which can sense when the RFID tag is away from the coat, and an alarm connected to the base module by a signal. The only connection by a signal mentioned in the patent is connection by a wire between the alarm and base module.

SmartyCoat has all the elements in a claim in the SmartCoat patent, except that where the SmartCoat patent showed only a wire connecting the alarm to the base module SmartyCoat uses bluetooth to wirelessly connect the alarm to the base module as the diagram on the facing page shows.

[As usual in these scenarios, this is simplified to keep focus on one issue at a time.]

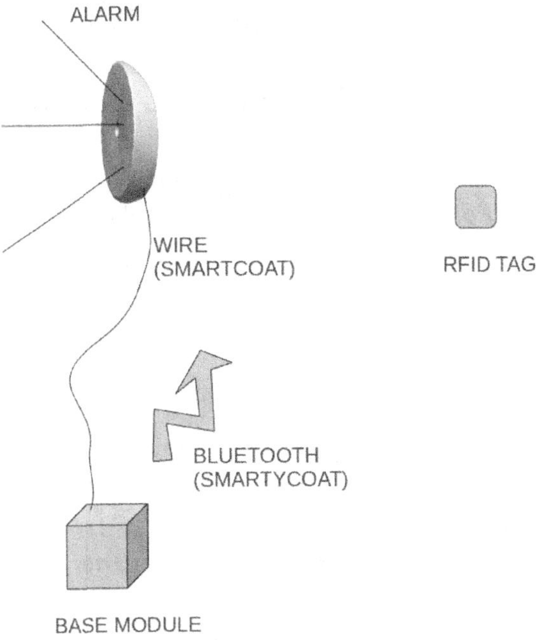

When an RFID tag is not within three feet of the base module, then the base module sends a signal to operate the alarm.

In this case **the jury must decide** if the bluetooth signal in SmartyCoat does, or does not:

1) Perform substantially the same *function* as the wire signal in SmartCoat. [We will take "substantially" to mean interchangeable]

2) In substantially the same *way*.

3) To yield substantially the same *result*.

[This "function – way – result" test is not at all clear cut. Patent trials are like this. The jury decision will hang on which side is most persuasive.]

Plaintiff's (QUBCO's) council and witness goals:

1) In the opening statement tell the jury that direct testimony by Brook will show that:

a) The SmartCoat patent is valid and is infringed by JUGCO.

b) The wire is preferred, but the blue tooth signal provides the same function the same way to give the same result.

2) In the direct testimony the plaintiff witness convinces the jury of the points in 1) above.

3) In the cross examination discredit Jessie and get Jessie to admit that the wire and bluetooth provide the same function the same way to give the same result.

4) In the closing statement retell all the points in 1), 2), and 3) above and ask the jury to find for the plaintiff.

Defendant's (JUGCO's) council and witness goals:

1) In the opening statement tell the jury that direct testimony by Jessie will show that:

a) SmartyCoat does not infringe SmartCoat.

b) Using bluetooth to wirelessly connect the alarm and the base module is not equivalent to using a wire because using bluetooth has surprising advantages such as reducing manufacturing costs.

2) In direct testimony Jessie convinces jury of the points in 1) above.

3) In cross examination discredit the plaintiff's witness and get plaintiff's witness to admit that that advantages of the wire are important.

4) In the closing statement retell all the points in 1), 2), and 3) above and ask the jury to find for the defendant.

[Remember that the attorneys and witnesses can not change the "facts" on the next page and in the depositions. The attorneys can not invent new "facts" either. Please review the sections of the general introduction which explain the scenarios and trials.]

[Also remember that attorneys can object if a witness is asked to testify about something not directly known by the witness and can object if a witness testifies about something not directly known by the witness.]

These "facts" are not disputed by plaintiff and defense and are entered as evidence: *[Y1 means year 1, Y2 year 2, Y3 year 3 – times are shorter than in reality for simplicity]*

Jan Y1: Quinn decides to invent SmartCoat to help friend Shelby keep from losing things. SmartCoat gives an alarm when RFID tagged things are not near the coat.

Feb Y1: Quinn and Brook incorporate as QUBCO. Brook works to perfect RFID used in SmartCoat.

Mar Y1: Shelby begins working at JUGCO.

Apr Y1: Jessie at JUGCO works on SmartyCoat.

Jan Y2: QUBCO makes and sells SmartCoat.

Feb Y2: JUGCO makes and sells SmartyCoat.

May Y2: QUBCO files patent application.

Jan Y3: QUBCO's patent granted

Feb Y3: QUBCO sues JUGCO for patent infringement.

The four depositions which follow are also "facts" which are entered as evidence.

[These "facts" are the raw material which the two attorney and witness teams use in opening statements, testimony, and closing statements.]

Deposition from the plaintiff's witness
Questioned by plaintiff's council
(Story for direct testimony)

[1] QUBCO Attorney: Please state your name and spell it for the record.
[2] Brook: Brook B R O O K.
[3] QUBCO Attorney: What is your occupation?
[4] Brook: When I was in high school I decided that I wanted to be an inventor so that's what I've done. I've made many inventions and innovations.
[5] QUBCO Attorney: What is the difference between an invention and an innovation?
[6] Brook: Inventions are what is claimed in a patent grant. Innovations are improvements which are valuable. Innovations are often inventions but not always. I'm good at perfecting inventions so that they become valuable innovations. That's why people hire me.
[7] QUBCO Attorney: Who do you work for now?
[8] Brook: I work with Quinn at QUBCO. He has lots of nifty ideas.
[9] QUBCO Attorney: But you get paid only in stock. Why do you work where you can only get paid in stock?
[10] Brook: I'm very good at what I do, but I'm no good at running a business. Since he was in high school Quinn always wanted to start a business and he's going to be very good at it – much better than me.
[11] QUBCO Attorney: Are you familiar with the RFID technology in QUBCO's SmartCoat?
[12] Brook: Yes, Quinn brought me in to perfect the idea. I made it so an alarm goes off when things are away from the coat.

[13] QUBCO Attorney: Was that new?
[14] Brook: Oh yes! We have an issued patent. We met all the requirements of the law. SmartCoat was not anticipated in any prior art.
[15] QUBCO Attorney: How does SmartCoat work exactly?
[16] Brook: There is a base module inside the coat that signals to an alarm when the RFID tag on an item is away from the base module.
[17] QUBCO Attorney: Where is the base module located?
[18] Brook: Usually in the coat pocket lining.
[19] QUBCO Attorney: Where is the alarm located?
[20] Brook: We put the alarm is underneath the collar so a wire to the base module can be hidden where no one can see it.
[21] QUBCO Attorney: Is there any other advantage in hiding the wire?
[22] Brook: Yes, there is less chance that it will be interfered with. I can make it more robust so that there is less chance that the alarm will fail to operate and so that there is less chance that spurious noise will operate the alarm.
[23] QUBCO Attorney: Are there other ways to connect the alarm and the base module?
[24] Brook: Yes, I could have made a wireless connection – could have used bluetooth like the SmartyCoat.
[25] QUBCO Attorney: Is there any difference in how bluetooth would function to operate the alarm?
[26] Brook: No, of course not, but it is not robust enough and too susceptible to spurious noise.

[27] QUBCO Attorney: Is there any difference in the way bluetooth would operate the alarm?
[28] Brook: No, of course not! The wire carries a signal to operate the alarm and bluetooth carries a signal to operate the alarm. A signal is a signal! The signal operates the alarm! No difference!
[29] QUBCO Attorney: Is there any difference in the result?
[30] Brook: No, of course not – the alarm is operated by the wire or by bluetooth. The wire is more reliable and less sensitive to spurious noise. Otherwise everything is the same.
[31] QUBCO Attorney: Thank you, no further questions.

[Remember to object when an attorney asks a witness to talk about anything the witness does not know directly or asks a witness to talk about issues other than the "facts" in evidence. And, object to answers when a witness says something not directly known by the witness or says something not about the "facts" in evidence.]

Deposition from the plaintiff's witness
Questioned by defendant's council
(Story for <u>cross examination</u>)

<u>[1] JUGCO Attorney</u> You stated that there is a wire from the alarm in SmartCoat collar to a base module. Correct? Or, not correct?
<u>[2] Brook:</u> Yes, there is.
<u>[3] JUGCO Attorney</u> What is the purpose of the wire?
<u>[4] Brook:</u> The wire is attached to the base module and the alarm. The wire carries a signal from the base module to the alarm when the base module does not detect the items anymore.
<u>[5] JUGCO Attorney</u> The wire is an important part of the SmartCoat. Correct? Or, not correct??
<u>[6] Brook:</u> Yes, it is important because it carries the signal to sound the alarm. That is whole purpose of the SmartCoat.
<u>[7] JUGCO Attorney</u> Thank you, no further questions.

Deposition from the defendant's witness
Questioned by defendant's council
(Story for direct testimony)

[1] JUGCO Attorney Please state your name and spell it for the record.
[2] Jessie: Jessie. J E S S I E.
[3] JUGCO Attorney What is your occupation?
[4] Jessie: I work for JUGCO.
[5] JUGCO Attorney What is JUGCO?
[6] Jessie: We were a third-generation, family-owned manufacturer of clothing – been around for nearly a hundred years. When competition from abroad got tough we changed focus to higher quality and more innovative goods, like SmartyCoat.
[7] JUGCO Attorney What is your job at JUGCO?
[8] Jessie: When we get an idea for an improved product, I work with all our other employees to devise enough improvements until we have a high quality innovation which does not infringe any patents and which will keep JUGCO going strong.
[9] JUGCO Attorney How do you do that?
[10] Jessie: We have several advantages. First,we have a very diverse bunch of employees, so that when we brainstorm we get many very interesting ideas. Second we know what works so we can quickly shelve ideas which won't work. Our late chief Emerson taught us "don't fall in love with your inventory. If it doesn't work send it to the rag factory right away." and we follow that rule.
[11] JUGCO Attorney Do you have much experience doing this?
[12] Jessie: Yes, we know the rules, we know that there are no shortcuts, we do the hard work to

perfect good ideas, we understand and manage the risks, we are very good at learning from others -- learning from patents -- and finding ways to make better products which do not infringe any patents.
[13] JUGCO Attorney Is it fair to learn from someone's patent and then go beyond it?
[14] Jessie: That's how patenting is intended to work. You get a right to exclude others in exchange for teaching some improvement. We learn from what the patent teaches and make the next improvement beyond what is taught in the patent. That's progress. That's what our Constitution promotes.
[15] JUGCO Attorney Are you familiar with the SmartCoat?
[16] Jessie: Yes, we learned about SmartCoat and devised ways around the patent to make SmartyCoat which is a higher quality and more innovative product.
[17] JUGCO Attorney What do you mean by "devised ways around the patent?"
[18] Jessie: As I said we find better ways which do not infringe patents.
[19] JUGCO Attorney How did you do that with SmartyCoat?
[20] Jessie: We don't infringe their patent if our alarm operates in a different way. By using bluetooth we operate the alarm in a very different way and we can place the alarm anywhere on the coat. That's a better idea.
[21] JUGCO Attorney How is bluetooth different from using a wire?
[22] Jessie: A wire limits where you can put the alarm. A wire requires extra manufacturing steps. A wire is messy.

[23] JUGCO Attorney Any other advantages?
[24] Jessie: Oh yes, it's much better. It is surprisingly cheaper because there are less manufacturing costs. And, now we have other high quality innovations also using bluetooth driven alarms.
25} JUGCO Attorney Thank you, no further questions.

Deposition from the defendant's witness
Questioned by plaintiff's council
(Story for <u>cross examination</u>)

<u>[1] QUBCO Attorney:</u> You stated that the JUGCO SmartyCoat technology uses bluetooth to communicate from the base module to the alarm. Correct? Or, not correct?
<u>[2] Jessie:</u> Yes, that's how it works.
<u>[3] QUBCO Attorney:</u> What happens when there is a bluetooth signal to the alarm?
<u>[4] Jessie:</u> The alarm sounds.
<u>[5] QUBCO Attorney:</u> How is sending a bluetooth signal different from sending a signal through a wire?
<u>[6] Jessie:</u> It is less expensive and you can put the alarm anywhere you want.
<u>[7] QUBCO Attorney:</u> A signal is a signal. Correct? Or, not correct?
<u>[8] Jessie:</u> Yes, but our signal is wireless.
<u>[9] QUBCO Attorney:</u> It functions the same and has the same result. Correct? Or, not correct?
<u>[10] Jessie:</u> Yes, but it is less expensive and you can put the alarm anywhere.
<u>[11] QUBCO Attorney:</u> Thank you, no further questions.

Can you identify some innovation wrongs which led to this trial?

Can you suggest some ways those innovation wrongs could have been put right?

If you were head of QUBCO what might you do to improve QUBCO's innovation rights?

If you were head of JUGCO, what might you do to improve JUGCO's innovation rights?

SmartCoat Novelty & Trade Secret Theft Trial

While QUBCO was making and selling SmartCoat and while their SmartCoat patent was pending. JUGCO was making and selling SmartyCoat, so QUBCO was eager to have their patent grant in order to sue JUGCO for infringement.

Surprise! QUBCO got an office action rejecting all the SmartCoat patent application claims because everything in the claims had been published in *Design News* by Jessie, a JUGCO employee, more than one year before QUBCO filed the SmartCoat patent application.

You can lose your patent right if your invention is made public more than one year before you apply an application for patent, even if someone steals your trade secrets and makes them public.

QUBCO's right to a patent was gone and there was no way to ever get it pack. Quinn was sure that Shelby had stolen information about SmartCoat and passed it to Jessie.

One way QUBCO could get some value from their invention of SmartCoat was to sue JUGCO for trade secret theft. So, QUBCO did sue JUGCO.

In our case **the jury must decide** if Shelby stole QUBCO's trade secrets and disclosed them to JUGCO and JUGCO benefited from the QUBCO trade secrets.

Here the burden is on the plaintiff to convince the jury that Shelby stole the trade secrets and disclosed them to JUGCO and JUGCO benefited from the QUBCO trade secrets. JUGCO in defense does not need to prove anything.

Plaintiff's (QUBCO's) council and witness goals:

1) In the opening statement tell the jury that direct testimony by Quinn will show that Shelby did steal QUBCO's trade secrets and did pass these secrets to Jessie.

2) In the direct testimony the plaintiff witness convinces the jury of the points in 1) above.

3) In the cross examination discredit Jessie and get Jessie to admit that:

a) Shelby had a direct path to pass QUBCO's trade secrets to Jessie.

b) There were several strange things about the JUGCO innovation process which all can be explained by Shelby passing on the stolen goods.

4) In the closing statement retell all the points in 1), 2), and 3) above and ask the jury to find for the plaintiff.

Defendant's (JUGCO's) council and witness goals:

1) In the opening statement tell the jury that direct testimony by Jessie will show that there is no evidence that JUGCO ever had any QUBCO trade secrets.

2) In direct testimony Jessie convinces the jury of the points in 1) above.

3) In cross examination discredit Quinn and get Quinn to admit that Quinn has no scientific evidence.

4) In the closing statement retell all the points in 1), 2), and 3) above and ask the jury to find for the defendant.

[Remember that the attorneys and witnesses can not change the "facts" on the next page and in the depositions. The attorneys can not invent new "facts" either. Please review the sections of the general introduction which explain the scenarios and trials.]

[Also remember that attorneys can object if a witness is asked to testify about something not directly known by the witness and can object if a witness testifies about something not directly known by the witness.]

These "facts" are not disputed by plaintiff and defense and are entered as evidence: *[Y1 means year 1, Y2 year 2, Y3 year 3 – times are shorter than in reality for simplicity]*

Jan Y1: Quinn decides to invent SmartCoat to help friend Shelby keep from losing things. SmartCoat gives an alarm when RFID tagged things are not near the coat.

Feb Y1: Quinn and Brook incorporate as QUBCO. Brook works to perfect RFID used in SmartCoat.

Mar Y1: Shelby begins working at JUGCO.

Apr Y1: Jessie at JUGCO works on SmartyCoat and publishes description in *Design News.*

Jan Y2: QUBCO makes and sells SmartCoat.

Feb Y2: JUGCO makes and sells SmartyCoat.

May Y2: QUBCO files patent application.

Jan Y3: QUBCO's patent granted

Feb Y3: QUBCO sues JUGCO for patent infringement.

The four depositions which follow are also "facts" which are entered as evidence.

[These "facts" are the raw material which the two attorney and witness teams use in opening statements, testimony, and closing statements.]

Deposition from the plaintiff's witness
Questioned by plaintiff's council
(Story for direct testimony)

[1] QUBCO Attorney: Please state your name and spell it for the record.
[2] Quinn: Quinn. Q U I N N.
[3] QUBCO Attorney: Why was your patent application for SmartCoat rejected?
[4] Quinn: Because the invention was disclosed in a publication by Jessie more than a year before our patent application was filed.
[5] QUBCO Attorney: How did it happen that your invention was disclosed by Jessie?
[6] Quinn: It happened because Shelby stole my trade secrets and passed them on to Jessie.
[7] QUBCO Attorney: Why do you think that your trade secrets were stolen?
[8] Quinn: I keep our trade secrets in a drawer in my desk with a big padlock on the drawer. One day the lock had been broken away and key papers were gone. The handle of my favorite hammer was broken to open that drawer.
[9] QUBCO Attorney: What was in those papers?
[10] Quinn: Those papers disclosed all the details of the invention. Those papers had the information the patent agent was using to write the patent application.
[11] QUBCO Attorney: Did you ever find those papers?
[12] Quinn: Yes, I found them on the copy machine and Shelby's dirty old glove was on the floor by the copy machine and the papers were all smudged with dirt from Shelby's glove.
[13] QUBCO Attorney: Thank you, no further questions.

Deposition from the plaintiff's witness
Questioned by defendant's council
(Story for cross examination)

[1] QUBCO Attorney: You did not have any scientific tests done to prove that it was Shelby's glove. Correct?
[2] Quinn: I know the glove well because Shelby was always forgetting it and I held it for Shelby many times.
[3] QUBCO Attorney: Yes or no please. Is it correct that you did not have tests done?
[4] Quinn: Yes.
[5] QUBCO Attorney: You did not do any linguistic tests to prove that Jessie's publication matched your papers. Correct?
[6] Quinn: Yes.
QUBCO Attorney: Thank you. No further questions.

Deposition from the defendant's witness
Questioned by defendant's council
(Story for direct testimony)

[1] JUGCO Attorney: Please state your name and spell it for the record.
[2] Jessie: Jesse. J E S S I E.
[3] JUGCO Attorney: Where did the ideas in your publication come from?
[4] Jessie: They are all ideas I put together.
[5] JUGCO Attorney: Did any of the ideas come from Shelby?
[6] Jessie: No.
[7] JUGCO Attorney: Thank you, no further questions.

[Remember that here the burden is on the plaintiff (Quinn) to prove that their trade secret was stolen. The defendant does not need to prove anything.]

Deposition from the defendant's witness
Questioned by plaintiff's council
(Story for cross examination)

[1] QUBCO Attorney: Do you know Shelby?
[2] Jessie: Yes.
[3] QUBCO Attorney: Shelby is a favorite of your boss. Correct?
[4] Jessie: Yes.
[5] QUBCO Attorney: When you work on a new product you invite suggestions from all JUGCO employees. Correct?
[6] Jessie: Yes.
[7] QUBCO Attorney: Your boss had Shelby process these and remove names so that you would not be biased by where the ideas came from. Correct?
[8] Jessie: Yes.
[9] QUBCO Attorney: Removing names was new. Correct?
[10] Jessie: Yes.
[11] QUBCO Attorney: You put these ideas from Shelby together for your publication. Correct?
[12] Jessie: Yes.
[13] QUBCO Attorney: You found it strange that the ideas fit together so well. Correct?
[14] Jessie: I don't remember.
[15] QUBCO Attorney: Shelby was actually submitting the ideas from QUBCO's trade secrets. Correct?
[16] Jessie: I don't know.
[17] QUBCO Attorney: Did any JUGCO employees get rewarded for contributions to SmartyCoat as usual?
[18] Jessie: No.

[19] QUBCO Attorney: Isn't that strange?
[20] Jessie: I don't . . .
[21] QUBCO Attorney: Thank you, no further questions.

Can you identify some innovation wrongs which led to this trial?

Can you suggest some ways those innovation wrongs could have been put right?

If you were head of QUBCO what might you do to improve QUBCO's innovation rights?

If you were head of JUGCO, what might you do to improve JUGCO's innovation rights?

SmartCoat Obvious Trial

Each of our set-ups for the trials here in Part One is different in order to illustrate different parts of the law.

This time, when QUBCO's SmartCoat patent issues and QUBCO sues JUGCO, JUGCO admits that they are infringing the SmartCoat patent but argues that the patent is invalid, that it never should have been allowed, because it is just an obvious change in well known art.

You have no patent right "if the differences between the subject matter sought to be patented and the prior art are such that the subject matter as a whole would have been obvious at the time the invention was made to a person having ordinary skill in the art to which said subject matter pertains" as the law states.

Deciding if a claim in a patent, or patent application, is obvious, or not, is perhaps the most slippery and burdensome part of patent practice.

In this case **the jury must decide** if the way SmartCoat detects when an RFID tag is away from SmartCoat:

1) was -- or was not -- one of known ways to do this, and thus obvious to try, by a person having ordinary skill in the art

2) was -- or was not -- expected to work by a person having ordinary skill in the art.

Slippery as this is, it provides a nice example of weighing testimony of dueling experts.

Plaintiff's (QUBCO's) council and witness goals:

1) In the <u>opening statement</u> tell the jury that direct testimony by Chris will show that:

a) The SmartCoat patent is valid and is infringed by JUGCO.
b) Chris has more than ordinary skill in the relevant art.
c) The way the SmartCoat tells when an RFID tag is away from the coat is not a way Chris would have tried and Chris is surprised that it works.

2) In the <u>direct testimony</u> Chris, the plaintiff's witness, convinces the jury of the points in 1) above.

3) In the <u>cross examination</u> discredit Pat's bragging and get Pat to admit that Pat just repairs things according to instructions in a

manual and is not qualified to testify about designs of circuits.

4) In the <u>closing statement</u> retell all the points in 1), 2), and 3) above and ask the jury to find for the plaintiff.

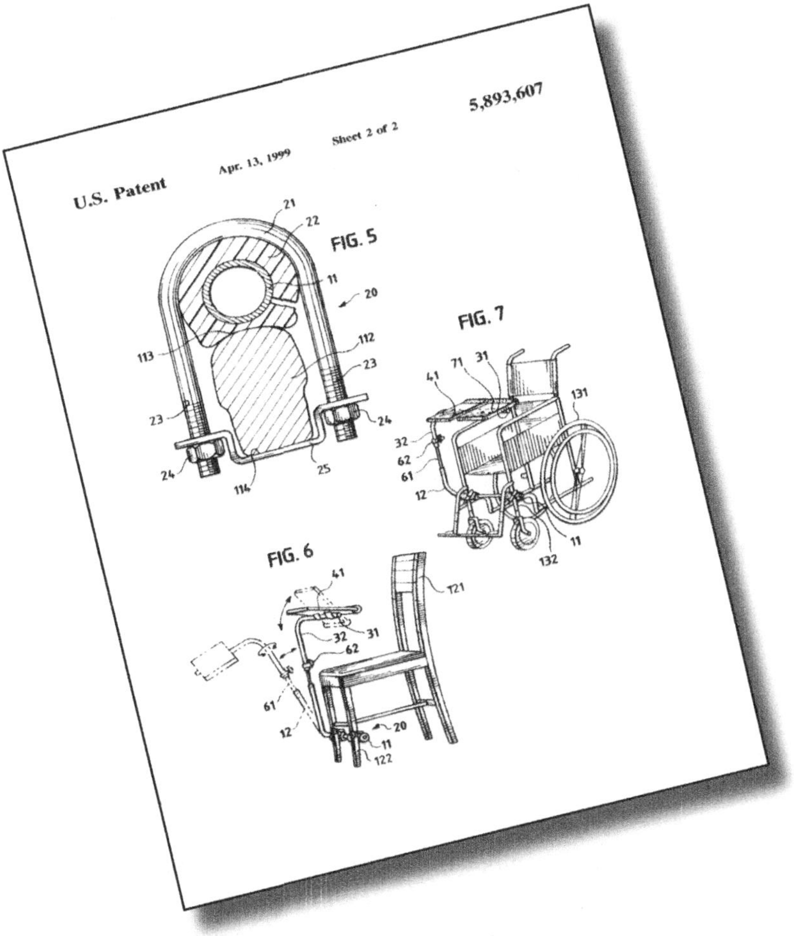

Defendant's (JUGCO's) council and witness goals:

1) In the opening statement tell the jury that direct testimony by Pat will show that:

a) The SmartCoat patent is not valid because the way SmartCoat tells when an RFID tag is far from the coat is obvious.
b) Pat has ordinary skill in the art.
c) Pat would have tried the way SmartCoat works and would have expected it to work.

2) In direct testimony Jessie convinces jury of the points in 1) above.

3) In cross examination discredit Chris and get Chris to admit that Chris is a biased witness because Chris is a neighbor and friend of Quinn and Brook.

4) In the closing statement retell all the points in 1), 2), and 3) above and ask the jury to find for the defendant.

[Remember that the attorneys and witnesses can not change the "facts" on the next page and in the depositions. The attorneys can not invent new "facts" either. Please review the sections of the general introduction which explain the scenarios and trials.]

[Also remember that attorneys can object if a witness is asked to testify about something not directly known by the witness and can object if a witness testifies about something not directly known by the witness.]

These "facts" are not disputed by plaintiff and defense and are entered as evidence: *[Y1 means year one – times are compressed much shorter than in reality for simplicity]*

Jan Y1: Quinn decides to invent SmartCoat to help friend Shelby keep from losing things. SmartCoat gives an alarm when RFID tagged things are not near the coat.

Feb Y1: Quinn and Brook incorporate as QUBCO. Brook perfects RFID used in SmartCoat.

Mar Y1: Shelby begins working at JUGCO.

Apr Y1: Jessie at JUGCO works on SmartyCoat.

May Y1: QUBCO files patent application.

Jun Y1: JUGCO sells SmartyCoats

Jul Y1: QUBCO's patent granted

Aug Y1: QUBCO sues JUGCO for patent infringement.

Sep Y1: Trial

The four depositions which follow are also "facts" which are entered as evidence.

[These "facts" are the raw material which the two attorney and witness teams use in opening statements, testimony, and closing statements.]

Deposition from the plaintiff's witness
Questioned by plaintiff's council
(Story for direct testimony)

[1] QUBCO Attorney: Please state your name and spell it for the record.
[2] Chris: Chris C H R I S.
[3] QUBCO Attorney: What is your occupation?
[4] Chris: I am an RFID expert.
[5] QUBCO Attorney: Are you a person having ordinary skill in RFID?
[6] Chris: No I have extraordinary skill. I'm a super expert. I have designed and built thousands of RFID security systems including for major retail chains.
[7] QUBCO Attorney: How do they work?
[8] Chris: When an RFID tag gets gets close to a sensor then an alarm is given so that security people can catch the thief and recover the stolen goods.
[9] QUBCO Attorney: How is SmartCoat different from this?
[10] Chris: In SmartCoat there is no alarm when the RFID tag is close to the coat. The alarm is given when the RFID tag is a pre-set distance from SmartCoat.
[11] QUBCO Attorney: Isn't that the same?
[12] Chris: No! It is much harder. Close is easy. Far is hard.
[13] QUBCO Attorney: Did you know about the work leading to Quinn's patent?
[14] Chris: Yes, their way of doing this is nifty. I would not have thought of it.
[15] Chris: You were surprised by their way to detect when an RFID tag is away from SmartCoat?

[16] Chris: Yes, I was surprised.
[17] QUBCO Attorney: Did you expect their way to detect when an RFID tag is away from SmartCoat to work?
[18] Chris: I thought that it would not work.
[19] QUBCO Attorney: In your expert opinion is this a discovery?
[20] Chris: Yes, in my super-expert opinion it is a discovery.
[21] QUBCO Attorney: In your expert opinion is this not obvious?
[22] Chris: Yes, in my super-expert opinion it is not obvious.
[23] QUBCO Attorney: Thank you. No further questions.

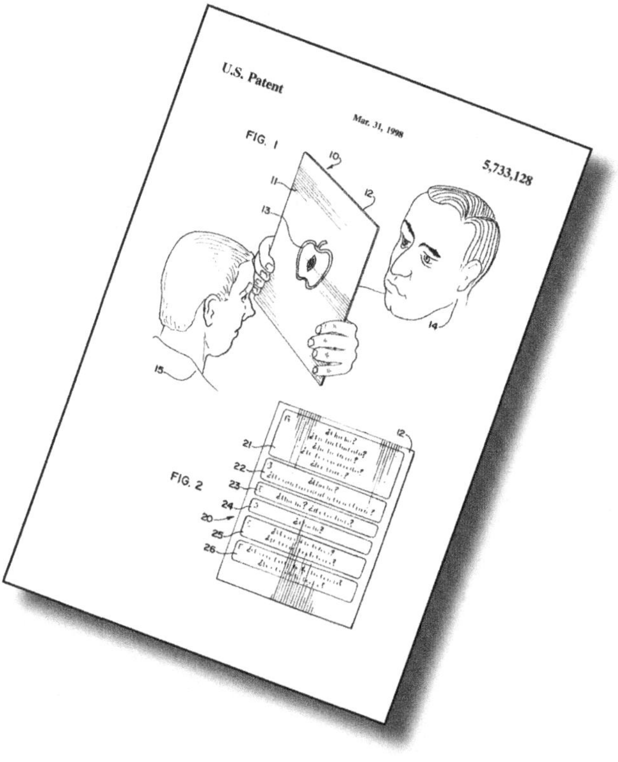

Deposition from the plaintiff's witness
Questioned by defendant's council
(Story for cross examination)

[1] JUGCO Attorney: Do you often sign the book where Quinn and Brook record their work?
[2] Chris: Yes, but I do not ever comment on their work or offer suggestions. I keep to my self any admiration for their discoveries.
[3] JUGCO Attorney: Does that make you an unofficial member of the SmartCoat team?
[4] Chris: No, I don't like SmartCoat. I live down the street from their shop so I witness things for them because I'm a super expert and a super-credible witness.
[5] JUGCO Attorney: Aren't you biased because of your familiarity with Quinn and Brook and their work.
[6] Chris: I am a super-credible witness. I am never biased. I don't even like Quinn and Brook. Brook is too much a know-it-all like many inventors. Quinn acts too much a big-shot like many start-up business people.
[7] JUGCO Attorney: Thank you. No further questions.

Deposition from the defendant's witness
Questioned by defendant's council
(Story for direct testimony)

[1] JUGCO Attorney: Please state your name and spell it for the record.
[2] Pat: Pat P A T.
[3] JUGCO Attorney: What is your occupation?
[4] Pat: I'm a security system repair technician.
[5] JUGCO Attorney: What is that?
[6] Pat: I repair security systems when they start malfunctioning. Like the people who repair your copy machines.
[7] JUGCO Attorney: Do you have ordinary skill in the art of RFID systems?
[8] Pat: I repair them so I must know how they work.
[9] JUGCO Attorney: Have you studied SmartCoat?
[10] Pat: Yes, it is a simple RFID application.
[11] JUGCO Attorney: Do you understand the way SmartCoat detects when an RFID tag is away from SmartCoat?
[12] Pat: Yes, it is very simple.
[13] JUGCO Attorney: In your opinion is the way SmartCoat detects when an RFID tag is away from SmartCoat something which you would have tried?
[14] Pat: Sure, I could have guessed that.
[15] JUGCO Attorney: Would you have expected it to work?
[16] Pat: Sure, it does work.
[17] JUGCO Attorney: In your opinion is the way SmartCoat detects when an RFID tag is away from SmartCoat obvious.

[18] Pat: Yes, I don't see any surprises. I don't see anything unexpected.
[19] JUGCO Attorney: Thank you. No further questions.

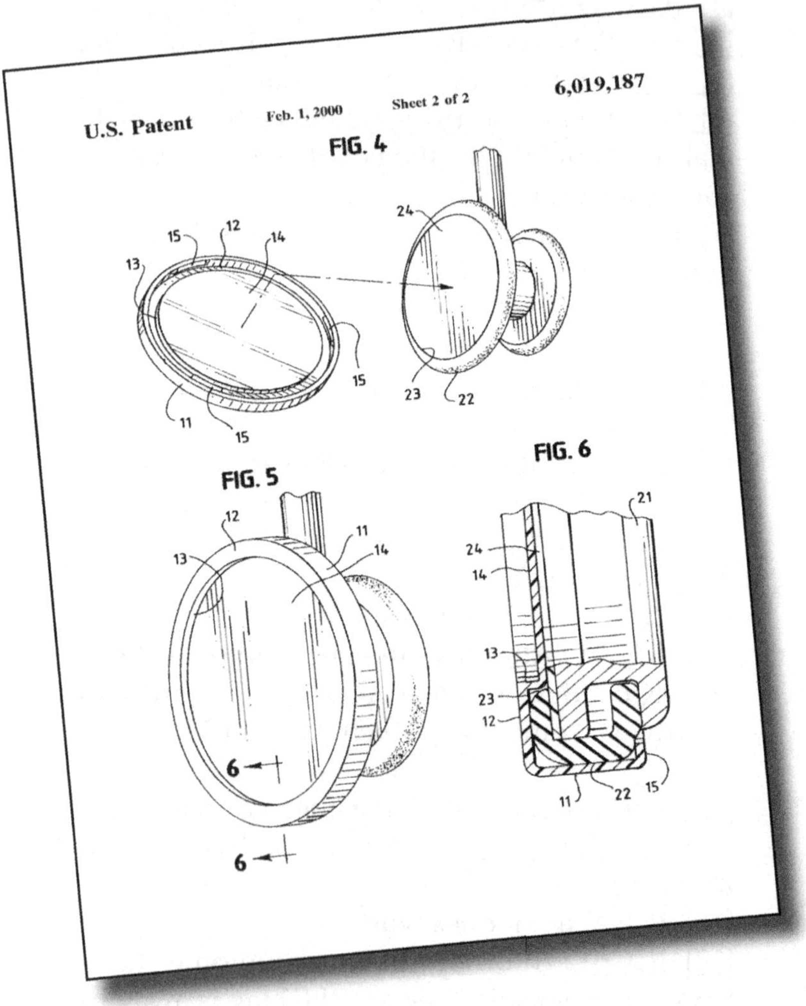

Deposition from the defendant's witness
Questioned by plaintiff's council
(Story for <u>cross examination</u>)

<u>[1] QUBCO Attorney:</u> You, said that you repair RFID systems, right?
<u>[2] Pat:</u> Yes, I went to technical school to learn how.
<u>[3] QUBCO Attorney:</u> Have you ever designed an RFID system?
<u>[4] Pat:</u> No, but I could.
<u>[5] QUBCO Attorney:</u> You follow a repair manual when you work on an RFID system. Correct?
<u>[6] Pat:</u> Yes, but I could do the work without the manual.
<u>[7] QUBCO Attorney:</u> Before you studied SmartCoat did you guess the way SmartCoat detects when an RFID tag is away from SmartCoat?
<u>[8] Pat:</u> No, I wasn't asked to guess.
<u>[9] QUBCO Attorney:</u> How many different RFID systems have you repaired?
<u>[10] Pat:</u> About one hundred.
<u>[11] QUBCO Attorney:</u> Did you ever see anything like the way SmartCoat detects when an RFID tag is away from SmartCoat in any of those one hundred?
<u>[12] Pat:</u> I don't remember.
<u>[13] QUBCO Attorney:</u> Thank you. No further questions.

Can you identify some innovation wrongs which led to this trial?

Can you suggest some ways those innovation wrongs could have been put right?

If you were head of QUBCO what might you do to improve QUBCO's innovation rights?

If you were head of JUGCO, what might you do to improve JUGCO's innovation rights?

SmartCoat Enable Trial

In this set-up, when QUBCO sues JUGCO for infringing QUBCO's SmartCoat patent by making and selling SmartyCoat, JUGCO admits that SmartyCoat does infringe the SmartCoat patent. In this trial JUGCO argues that the SmartCoat patent is not valid because the patent specification does not meet the enable requirement of the law.

The law requires that your patent specification must "contain a written description of the invention, and of the manner and process of making and using it, in such full, clear, concise, and exact terms as to **enable** any person skilled in the art to which it pertains, or with which it is most nearly connected, to make and use the same"

In this case **the jury must decide** if a person having ordinary skill in RFID art would -- or would not -- be able understand how SmartCoat detects when an RFID tag is away from SmartCoat.

This trial is going to look like a funny reflection of the obvious trial.

Plaintiff's (QUBCO's) council and witness goals:

1) In the opening statement tell the jury that direct testimony by Brook will show that:

a) The SmartCoat patent is valid and is infringed by JUGCO.

b) There is nothing left out of the patent specification except things which are obvious to skilled persons.

2) In the direct testimony Brook, the plaintiff's witness, convinces the jury of the points in 1) above.

3) In the cross examination discredit Jessie and get Jessie to admit that:

a) JUGCO depended on JUGCO employees and they are not skilled in electric circuits.

b) If JUGCO had consulted someone skilled in electric circuits there would have been no trouble.

4) In the closing statement retell all the points in 1), 2), and 3) above and ask the jury to find for the plaintiff.

Defendant's (JUGCO's) council and witness goals:

1) In the opening statement tell the jury that direct testimony by Jessie will show that:

a) The SmartCoat patent is not valid because the specification does not enable the way the base module tells when an RFID tag is away from the coat.

b) It took JUGCO many experiments to learn how to do this.

2) In the direct testimony Jessy convinces jury of the points in 1) above.

3) In the cross examination discredit Brook and get Brook to admit that Brook might have unconsciously invented a way to tell when an RFID tag is away from the coat, which should have been disclosed.

4) In the closing statement retell all the points in 1), 2), and 3) above and ask the jury to find for the defendant.

[Remember that the attorneys and witnesses can not change the "facts" on the next page and in the depositions. The attorneys can not invent new "facts" either. Please review the sections of the general introduction which explain the scenarios and trials.]

[Also remember that attorneys can object if a witness is asked to testify about something not directly known by the witness and can object if a witness testifies about something not directly known by the witness.]

These "facts" are not disputed by plaintiff and defense and are entered as evidence: *[Y1 means year one – times are compressed much shorter than in reality for simplicity]*

Jan Y1: Quinn decides to invent SmartCoat to help friend Shelby keep from losing things. SmartCoat gives an alarm when RFID tagged things are not near the coat.

Feb Y1: Quinn and Brook incorporate as QUBCO. Brook perfects RFID used in SmartCoat.

Mar Y1: Shelby begins working at JUGCO.

Apr Y1: Jessie at JUGCO works on SmartyCoat.

May Y1: QUBCO files patent application.

Jun Y1: JUGCO sells SmartyCoats.

Jul Y1: QUBCO's patent granted.

Aug Y1: QUBCO sues JUGCO for patent infringement.

Sep Y1: Trial.

The four depositions which follow are also "facts" which are entered as evidence.

[These "facts" are the raw material which the two attorney and witness teams use in opening statements, testimony, and closing statements.]

Deposition from the plaintiff's witness
Questioned by plaintiff's council
(Story for direct testimony)

[1] QUBCO Attorney: Please state your name and spell it for the record.
[2] Brook: Brook B R O O K.
[3] QUBCO Attorney: What is your occupation?
[4] Brook: I'm an inventor. I work for Quinn.
[5] QUBCO Attorney: Did you work on SmartCoat?
[6] Brook: Yes, I did it was my main job for a while.
[7] QUBCO Attorney: Did you help with the patent application for SmartCoat?
[8] Brook: Yes, I did. I worked with the patent agents. They did great work.
[9] QUBCO Attorney: Why was the way SmartCoat detects when an RFID tag is away from SmartCoat not described in detail?
[10] Brook: I told the patent agents that this was a minor and obvious engineering detail which did not need to be disclosed, no more that one need disclose how to tighten a nut on a bolt to fasten two parts of a device together.
[11] QUBCO Attorney: How did you find the the way SmartCoat detects when an RFID tag is away from SmartCoat?
[12] Brook: I don't remember clearly. I was mainly helping Quinn with his business plan and was working on that simple detail while eating lunches.
[13] QUBCO Attorney: This step did not concern you?
[14] Brook: No, there was no surprise, just ordinary engineering I could do in my sleep.

[15] QUBCO Attorney: Thank you. No further questions.

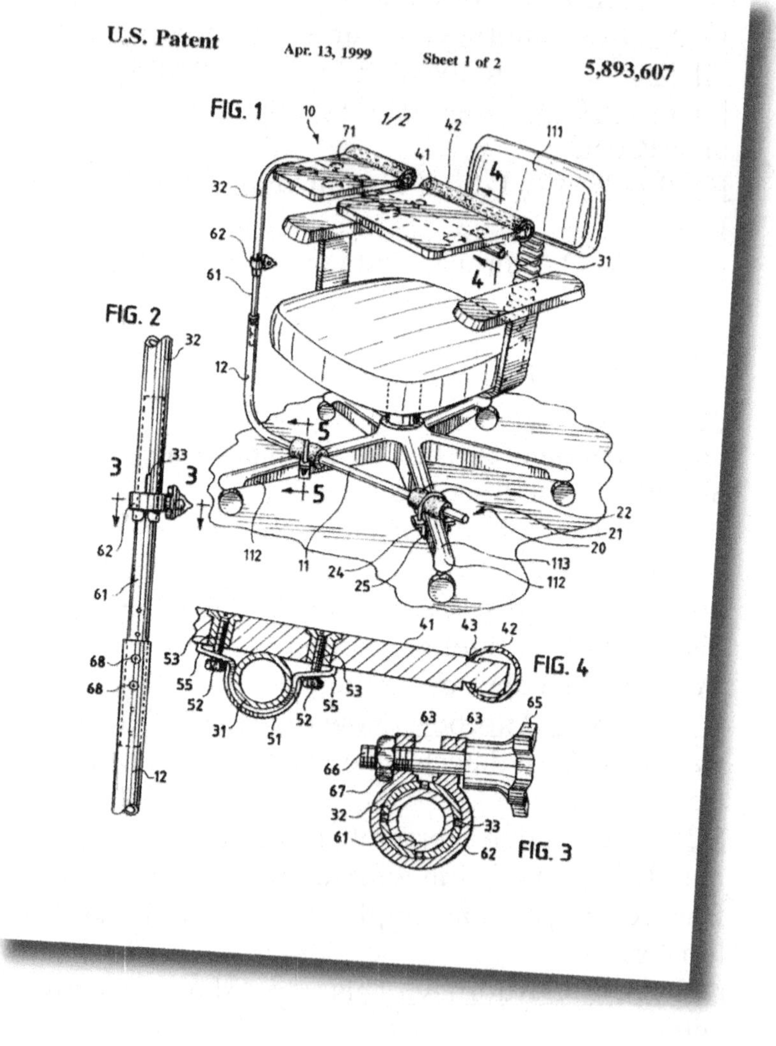

Deposition from the plaintiff's witness
Questioned by defendant's council
(Story for cross examination)

[1] JUGCO Attorney: You said that you were not paying attention to the way SmartCoat detects when an RFID tag is away from SmartCoat. Correct?
[2] Brook: Yes, it is just ordinary engineering.
[3] JUGCO Attorney: You were not systematically reviewing known engineering ways for detecting RFID tags Correct?
[4] Brook: Yes, there was no need to.
[5] JUGCO Attorney: So, one day the answer just came to you. Correct?
[6] Brook: Yes, no surprise.
[7] JUGCO Attorney: You made many inventions in the past. Correct?
[8] Brook: Yes, I'm very good at that.
[9] JUGCO Attorney: Your inventions were not just obvious variations on well known things. Correct?
[10] Brook: Yes, otherwise they would not be real inventions which are patented.
[11] JUGCO Attorney: You have a very impressive list of patented inventions. Did some of them just come to you?
[12] Brook: Sometimes it works that way.
[13] JUGCO Attorney: So, the way you found for SmartCoat to detect when an RFID tag is away from SmartCoat could also have been a real invention which just came to you. Correct?
[14] Brook: Yes, but . . .
[15] JUGCO Attorney: Thank you. No further questions.

Deposition from the defendant's witness
Questioned by defendant's council
(Story for direct testimony)

[1] JUGCO Attorney: Please state your name and spell it for the record.
[2] Jessie: Jessie J E S S I E.
[3] JUGCO Attorney: What is your occupation?
[4] Jessie: I work for JUGCO.
[5] JUGCO Attorney: What was your response to SmartCoat?
[6] Jessie: First, we took it apart to discover how it worked.
[7] JUGCO Attorney: Any problems there?
[8] Jessie: Yes, the circuits were carefully disguised so we could not find out how SmartCoat detected when an RFID tag is away from SmartCoat.
[9] JUGCO Attorney: What did you do?
[10] Jessie: We followed our normal brainstorming-for-innovation process where we get ideas from all of our employees for a way for our SmartyCoat to detect when an RFID tag is away from SmartyCoat.
[11] JUGCO Attorney: How did that go?
[12] Jessie: It was torture. We did many experiments with no luck. Then on day a spacey shipping clerk, who's a big fan of science fiction and other weird stuff, brought an idea I gave it to our maintenance people. They said that it was nonsense and wouldn't work
[13] JUGCO Attorney: So what did you do then?
[14] Jessie: For our innovation process to work we must treat all ideas with full respect and try them out. We tried it and it worked. Big surprise. Our maintenance people were dumbfounded.

[15] JUGCO Attorney: What did you do when Quinn sued you for infringement.
[16] Jessie: We studied Quinn's patent and we found that the patent does not disclose how SmartCoat detects when an RFID tag is away from SmartCoat. Since in our experience this took great experimenting and still the answer is a surprise we knew that Quinn's patent is not valid.
[17] JUGCO Attorney: Thank you. No further questions.

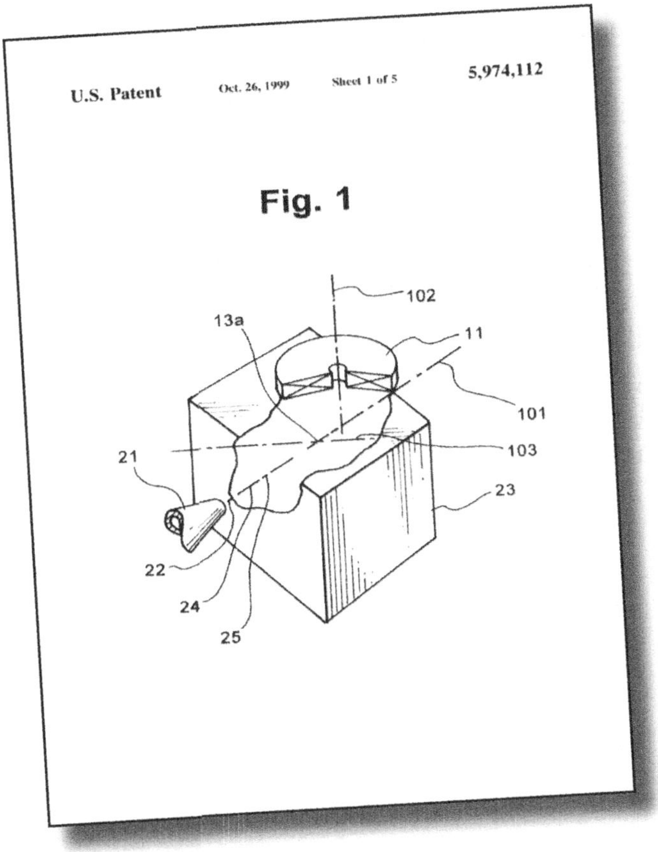

Deposition from the defendant's witness
Questioned by plaintiff's council
(Story for cross examination)

[1] QUBCO Attorney: You said that in your brainstorming-for-innovation process at JUGCO you give every idea a fair chance and try them all out. Correct?
[2] Jessie: Right, that process has served us very well.
[3] QUBCO Attorney: You test every idea, even ideas which seem crazy. Correct?
[4] Jessie: Yes, that's the best way to get the best ideas.
[5] QUBCO Attorney: You do this so that everyone feels empowered to submit ideas. Correct?
[6] Jessie: That is correct. It works.
[7] QUBCO Attorney: But sometimes you could find an obvious answer quicker if you just followed ordinary engineering practice. Correct?
[8] Jessie: Yes, but that . . .
QUBCO Attorney: Thank you. No further questions.

Can you identify some innovation wrongs which led to this trial?

Can you suggest some ways those innovation wrongs could have been put right?

If you were head of QUBCO what might you do to improve QUBCO's innovation rights?

If you were head of JUGCO, what might you do to improve JUGCO's innovation rights?

SmartCoat License Arbitration

In this scenario when QUBCO's SmartCoat patent issues, rather than sue JUGCO, QUBCO offers to give JUGCO a license to SmartCoat so that JUGCO could continue to make and sell SmartyCoat and pay QUBCO a royalty on each SmartyCoat sold.

The license specified that the license was limited to the invention described and claimed in the issued patent. The license also specified that any dispute would be settled by arbitration and further specified that the arbitrator hold a trial like our mock trials.

Arbitration is a proceeding in which a dispute is resolved outside of courts by an impartial adjudicator whose decision the parties to the dispute have agreed will be final and binding. That is a very much simplified statement. It's a good idea to learn more about arbitration and mediation, which are much less costly than courts and are much better ways to right wrongs.

Soon QUBCO was making and selling SmartCoats with a removable wire connecting the alarm and the base module. JUGCO also started making and selling SmartyCoats with a removable wire. QUBCO invoked the arbitration clause in the license to stop JUGCO selling coats with removable wires.

JUGCO responded that since a removable wire was the "best mode contemplated by the inventor of carrying out his invention" and was required by the patent law to be in the patent application, the removable wire is therefore part of what is covered by the license.

In our case **the arbitrator must decide** if the removable wire was, or was not, thought to be the best idea at the time Quinn's patent application was filed. This is especially slippery because it is about intent.

Plaintiff's (QUBCO's) council and witness goals:

1) In the opening statement tell the jury that direct testimony by Brook will show

a) That the idea for a removable wire came after the patent application was filed.
b) That Brook is the inventor and did not contemplate using a removable wire until after the application was filed.

2) In the direct testimony Brook, the plaintiff witness, convinces the jury of the points in 1) above.

3) In the cross examination discredit Shelby and get Shelby to admit that the only evidence is from Shelby and Shelby has a grudge against Quinn:

4) In the closing statement retell all the points in 1), 2), and 3) above and ask the jury to find for the plaintiff.

Defendant's (JUGCO's) council and witness goals:

1) In the opening statement tell the jury that direct testimony by Shelby will show that Quinn thought that a removable wire was best before the patent application was filed:

2) In direct testimony Shelby, the defense witness, convinces jury of the points in 1) above.

3) In cross examination discredit Brook and get Brook to admit that Quinn was also an inventor of SmartCoat and Quinn did think that a removable wire was best before the patent application was filed.

4) In the closing statement retell all the points in 1), 2), and 3) above and ask the jury to find for the defendant.

[Remember that the attorneys and witnesses can not change the "facts" on the next page and in the depositions. The attorneys can not invent new "facts" either. Please review the sections of the general introduction which explain the scenarios and trials.]

[Also remember that attorneys can object if a witness is asked to testify about something not directly known by the witness and can object if a witness testifies about something not directly known by the witness.]

These "facts" are not disputed by plaintiff and defense and are entered as evidence: *[Y1 means year 1, Y2 year 2, Y3 year 3 – times are shorter than in reality for simplicity]*

Jan Y1: Quinn decides to invent SmartCoat to help friend Shelby keep from losing things. SmartCoat gives an alarm when RFID tagged things are not near the coat.

Feb Y1: Quinn and Brook incorporate as QUBCO. Brook works to perfect RFID used in SmartCoat.

Mar Y1: Shelby begins working at JUGCO.

Apr Y1: Jessie at JUGCO works on SmartyCoat.

Jan Y2: QUBCO makes and sells SmartCoat.

Feb Y2: JUGCO makes and sells SmartyCoat.

May Y2: QUBCO files patent application.

Jan Y3: QUBCO's patent granted and QUBCO licenses JUGCO

Feb Y3: QUBCO invokes arbitration clause.

The four depositions which follow are also "facts" which are entered as evidence.

[These "facts" are the raw material which the two attorney and witness teams use in opening statements, testimony, and closing statements.]

Deposition from the plaintiff's witness
Questioned by plaintiff's council
(Story for direct testimony)

[1] QUBCO Attorney: Please state your name and spell it for the record.
[2] Brook: Brook B R O O K
[3] QUBCO Attorney: You are listed as inventor on Quinn's patent for SmartCoat. Please explain.
[4] Brook: I invented SmartCoat. I was hired by Quinn to invent SmartCoat. I worked with the patent attorney to write the patent application.
[5] QUBCO Attorney: The patent does not mention a removable wire between the base module and the alarm. Why is this not there?
[6] Brook: It was not part of my invention at that time.
[7] QUBCO Attorney: When did you switch to the removable wire?
[8] Brook: It was soon after the patent application was filed. I tested how SmartCoat was effected by cleaning and found that the permanent wire I was using was hurt by the cleaning.
[9] QUBCO Attorney: What did you do then?
[10] Brook: Quinn told me to make the wire removable for cleaning. He's the boss, so I did.
[11] QUBCO Attorney: So the removable wire came after the patent application was filed?
[12] Brook: Yes, it was not part of my invention when the application was filed.
[13] QUBCO Attorney: Thank you. No further questions.

Deposition from the plaintiff's witness
Questioned by defendant's council
(Story for cross examination)

[1] JUGCO Attorney: You said that the removable wire was not part of your invention when the patent application was filed. Correct?
[2] Brook: Yes.
[3] JUGCO Attorney: Quinn is also listed as inventor also. Correct?
[4] Brook: That's a technicality because some of the ideas were originally his, but I made it work.
[5] JUGCO Attorney: Yes or no please.
[6] Brook: Yes.
[7] JUGCO Attorney: Quinn told you to try a removable wire before the the patent application was filed. Correct?
[8] Brook: No.
[9] JUGCO Attorney: Quinn thought a removable wire would be best. Correct?
[10] Brook: I don't know.
[11] JUGCO Attorney: You are paid in stock. Correct?
[12] Brook: Yes.
[13] JUGCO Attorney: If Quinn loses this case the stock will be worth very little. Correct?
[14] Brook: I don't know.
[15] JUGCO Attorney: Thank you. No further questions.

Deposition from the defendant's witness
Questioned by defendant's council
(Story for direct testimony)

[1] JUGCO Attorney: Please state your name and spell it for the record.
[2] Shelby: Shelby S H E L B Y.
[3] JUGCO Attorney: What is your connection with SmartCoat?
[4] Shelby: Quinn decided to invent SmartCoat because I lose things.
[5] JUGCO Attorney: Did you contribute any ideas to SmartCoat?
[6] Shelby: Early on I told Quinn that the wire connecting the base module and the alarm should be removable for cleaning.
[7] JUGCO Attorney: What did he say?
[8] Shelby: He said, "yes, that would be best."
[9] JUGCO Attorney: What happened?
[10] Shelby: Brook vetoed the idea: After that I left to work for JUGCO.
[11] JUGCO Attorney: Thank you. No further questions.

Deposition from the defendant's witness
Questioned by plaintiff's council
(Story for cross examination)

[1] QUBCO Attorney: You left to work at JUGCO because Quinn paid more attention to Brook than to you. Correct?
[2] Shelby: I left for many reasons.
[3] QUBCO Attorney: But, you were angry with Quinn. Correct?
[4] Shelby: He's a jerk!
[5] QUBCO Attorney: So you come here to tell the arbitrator that Quinn said that a removable wire would be best in order to punish Quinn for paying more attention to Brook than to you. Correct?
[6] Shelby: I . . .
[7] QUBCO Attorney: Thank you. No further questions.

Can you identify some innovation wrongs which led to this trial?

Can you suggest some ways those innovation wrongs could have been put right?

If you were head of QUBCO what might you do to improve QUBCO's innovation rights?

If you were head of JUGCO, what might you do to improve JUGCO's innovation rights?

SmartCoat On Sale Trial

In this scenario, when QUBCO sues JUGCO because SmartyCoat infringes QUBCO's SmartCoat patent, JUGCO argues that QUBCO's SmartCoat patent is not valid because QUBCO offered SmartCoat for sale more than a year prior to filing the patent application.

The law says that you can lose your patent right if you do not file a patent application before you sell or offer to sell the invention. There are two requirements:

1) A commercial sale or offer for sale.

2) The invention was ready for patenting at the time of sale or offer for sale.

Shelby claims that Quinn said Shelby could have the first SmartCoat early in year one. And, Shelby claims that this offer was confirmed when Shelby confronted Brook in April of year one.

If Quinn did make an offer for sale early in year one it would not count because SmartCoat was not ready for patenting then. If there was a confirmation of an earlier offer for sale in April of year one (when SmartCoat was ready for patenting), then there would be an offer for sale as of the earlier date.

So, in this case **the jury must decide** if Shelby's confrontation with Brook was a confirmation of an earlier offer for sale.

[For simplicity we can ignore issues related to determining if an offer for sale is a "commercial" offer for sale.]

Plaintiff's (QUBCO's) council and witness goals:

1) In the opening statement tell the jury that direct testimony by Brook will show that:

a) The SmartCoat patent is valid and is infringed by JUGCO.

b) There was no confirmation of an earlier offer for sale.

c) Shelby forfeited any right to a SmartCoat.

d) Shelby was told that Shelby could buy a SmartCoat only when SmartCoats were on sale for everyone.

2) In the direct testimony Brook, the plaintiff's witness, convinces the jury of the points in 1) above.

3) In the cross examination discredit Shelby and get Shelby to admit that Shelby can't remember what Brook said.

4) In the closing statement retell all the points in 1), 2), and 3) above and ask the jury to find for the plaintiff.

Defendant's (JUGCO's) council and witness goals:

1) In the opening statement tell the jury that direct testimony by Shelby will show that:

a) Quinn offered a SmartCoat to Shelby early in year one;

b) Brook, speaking for Quinn, confirmed that offer in April of year one.

2) In the direct testimony Jessie convinces jury of the points in 1) above.

3) In cross examination discredit Brook and get Brook to admit that Brook did not tell Shelby that the earlier offer was forfeited.

4) In the closing statement retell all the points in 1), 2), and 3) above and ask the jury to find for the defendant.

[Remember that the attorneys and witnesses can not change the "facts" on the next page and in the depositions. The attorneys can not invent new "facts" either. Please review the sections of the general introduction which explain the scenarios and trials.]

[Also remember that attorneys can object if a witness is asked to testify about something not directly known by the witness and can object if a witness testifies about something not directly known by the witness.]

These "facts" are not disputed by plaintiff and defense and are entered as evidence: *[Y1 means year 1, Y2 year 2, Y3 year 3 – times are shorter than in reality for simplicity]*

Jan Y1: Quinn decides to invent SmartCoat to help friend Shelby keep from losing things. SmartCoat gives an alarm when RFID tagged things are not near the coat.

Feb Y1: Quinn and Brook incorporate as QUBCO. Brook works to perfect RFID used in SmartCoat.

Mar Y1: Shelby begins working at JUGCO.

Apr Y1: Jessie at JUGCO works on SmartyCoat.

Jan Y2: QUBCO makes and sells SmartCoat.

Feb Y2: JUGCO makes and sells SmartyCoat.

May Y2: QUBCO files patent application.

Jan Y3: QUBCO's patent granted

Feb Y3: QUBCO sues JUGCO for patent infringement.

The four depositions which follow are also "facts" which are entered as evidence.

[These "facts" are the raw material which the two attorney and witness teams use in opening statements, testimony, and closing statements.]

Deposition from the plaintiff's witness
Questioned by plaintiff's council
(Story for direct testimony)

[1] QUBCO Attorney: Please state your name and spell it for us.
[2] Brook: Brook B R O O K.
[3] QUBCO Attorney: What is your occupation?
[4] Brook: When I was in high school I decided that I wanted to be an inventor so that's what I've done. I've made many inventions and innovations. I work for Quinn. He has lots of nifty ideas.
[5] QUBCO Attorney: But he can only pay you in stock. Why do you work for someone who can only pay you in stock?
[6] Brook: I'm very good at what I do, but I'm no good at running a business. Since he was in high school Quinn always wanted to start a business and he's going to be very good at it – much better than me.
[7] QUBCO Attorney: Are you familiar with Quinn's SmartCoat?
[8] Brook: Yes, Quinn hired me to perfect the idea. I made it so an alarm goes off when things are away from the coat.
[9] QUBCO Attorney: Was that new?
[10] Brook: Oh yes! We have an issued patent. We met all the requirements of the law. SmartCoat was not anticipated in any prior art.
[11] QUBCO Attorney: Tell us what happened when Shelby came to see you?
[12] Brook: Shelby came by and asked me for a SmartCoat. I told her that I would not give her one. She said that Quinn promissed her the first SmartCoat.

[13] QUBCO Attorney: What did you do then?
[14] Brook: I went to where Quinn was working. He refused to talk to Shelby. He said that she forfeited any claim to a SmartCoat when Shelby went to work at JUGCO. He told me to tell her that she could buy a SmartCoat like anyone else.
[15] QUBCO Attorney: What did Shelby do when you told her this?
[16] Brook: Shelby stomped away muttering something about Quinn.
[17] QUBCO Attorney: Thank you. No furtther questions.

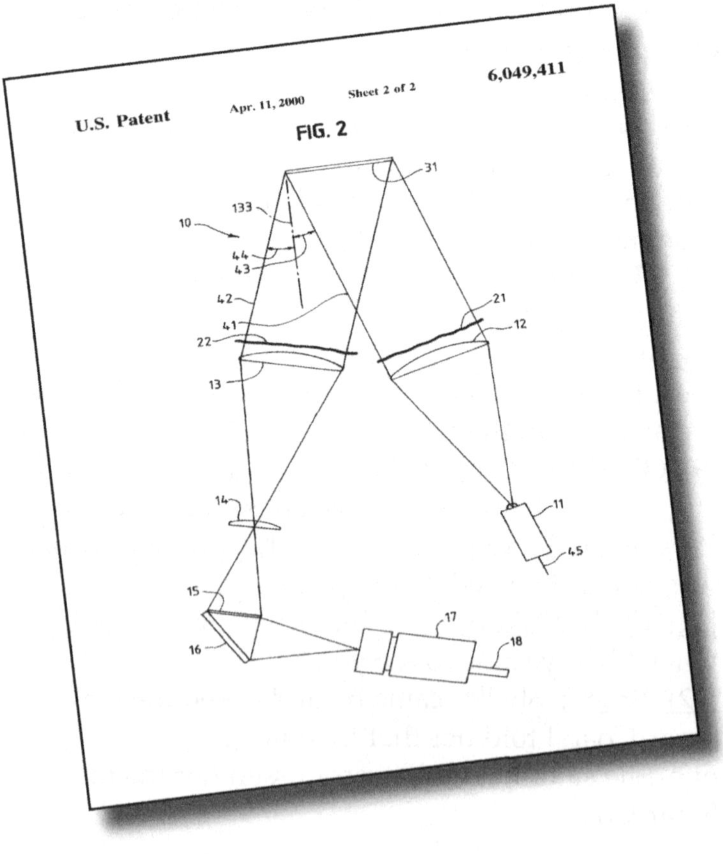

Deposition from the plaintiff's witness
Questioned by defendant's council
(Story for cross examination)

[1] JUGCO Attorney: When you told Shelby that she could buy a SmartCoat Shelby said that she would hold Quinn to that offer. Correct?
[2] Brook: Shelby muttered something that I did not hear.
[3] JUGCO Attorney: She could have muttered that she would hold Quinn to the offer for sale. Correct?
[4] Brook: I don't know.
[5] JUGCO Attorney: You did not say anything to Shelby to indicate that the offer for sale was not a genuine offer for sale. Correct?
[6] Brook: I don't know what you mean by a "genuine offer for sale" I only told Shelby what Quinn asked me to tell Shelby.
[7] JUGCO Attorney: Thank you. No further questions.

Deposition from the defendant's witness
Questioned by defendant's council
(Story for direct testimony)

[1] JUGCO Attorney: Please tell us your name and spell it for us.
[2] Shelby: Shelby S H E L B Y.
[3] JUGCO Attorney: What is your occupation?
[4] Shelby: I work for JUGCO.
[5] JUGCO Attorney: What is JUGCO?
[6] Shelby: We're a third-generation, family-owned manufacturer of clothing – been around for nearly a hundred years. When competition from abroad got tough we changed focus to higher quality and more innovative goods, like SmartyCoat.
[7] JUGCO Attorney: What is your job at JUGCO?
[8] Shelby: I scout for new product ideas.
[9] JUGCO Attorney: What do you know about SmartCoat?
[10] Shelby: Quinn decided to invent SmartCoat to help me keep from loosing things. Quinn promised me that I would have the first SmartCoat.
[11] JUGCO Attorney: Why did you go to work at JUGCO?
[12] Shelby: I wanted to work for a real company actually making and selling things, actually bringing new products to market. I got tired of Quinn's big ideas and little action.
[13] JUGCO Attorney: What happened when you went to claim the SmartCoat which Quinn promised you?
[14] Shelby: Quinn refused to see me. Brook told me that Quinn said that I could buy a SmartCoat. I

said that I will hold the jerk Quinn to that offer for sale.
[15] JUGCO Attorney: Did Brook say anything to indicate that this was not a genuine offer for sale?
[16] Shelby: No. Brook said nothing like that.
[17] JUGCO Attorney: Thank you. No further questions.

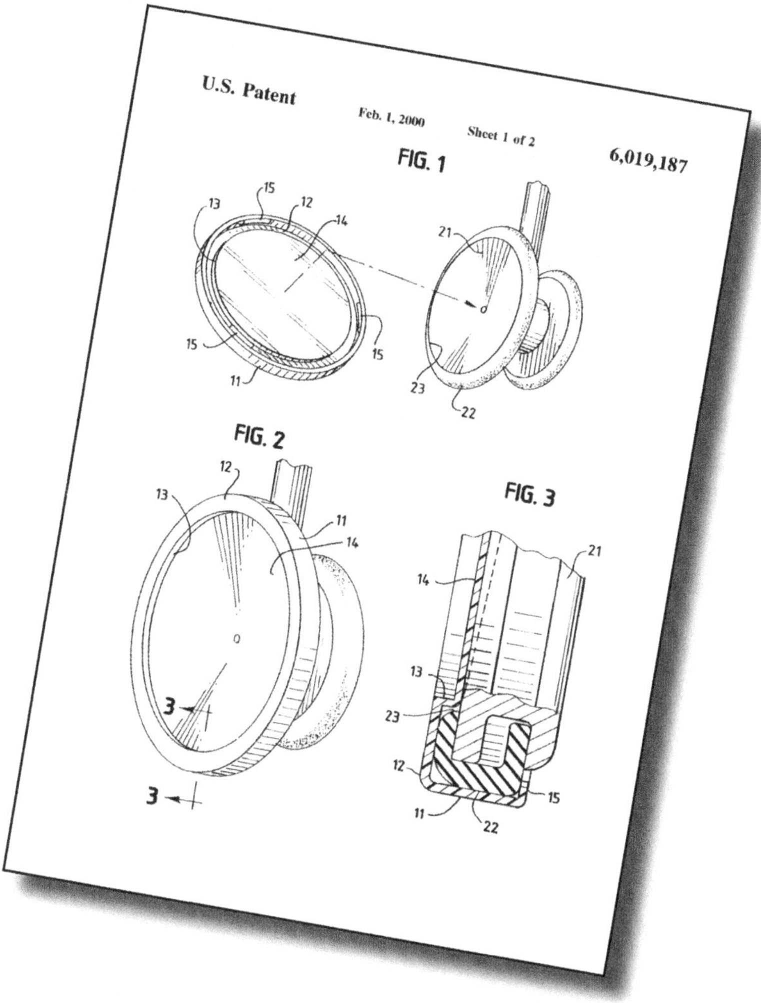

Deposition from the defendant's witness
Questioned by plaintiff's council
(Story for cross examination)

[1] QUBCO Attorney: Brook told you that Quinn said that you could buy a SmartCoat like anyone else. Correct?
[2] Shelby: Brook said that I could buy a SmartCoat.
[3] QUBCO Attorney: Yes or no please.
[4] Shelby: I don't remember.
[5] QUBCO Attorney: When Brook told you that you could buy a SmartCoat like anyone else that meant that you could buy a SmartCoat when SmartCoats were on sale for everyone. Correct?
[6] Shelby: I don't remember Brook saying that.
[7] QUBCO Attorney: Saying that you could buy a SmartCoat like anyone else meant that you could buy a SmartCoat when SmartCoats were on sale for everyone. Correct?
[8] Shelby: I don't know.
[9] QUBCO Attorney: Thank you. No further questions.

Can you identify some innovation wrongs which led to this trial?

Can you suggest some ways those innovation wrongs could have been put right?

If you were head of QUBCO what might you do to improve QUBCO's innovation rights?

If you were head of JUGCO, what might you do to improve JUGCO's innovation rights?

SmartCoat Employer Rights Trial

Before Quinn decided to start a smart products business Quinn worked at a JUGCO warehouse. In this scenario, when QUBCO sues JUGCO because SmartyCoat infringes QUBCO's SmartCoat patent, JUGCO argues that since Quinn had worked for JUGCO Quinn is obliged to give the invention rights to JUGCO.

The Illinois Employment Patent Act lists conditions when an employee is obliged to assign an invention to an employer.

Since Quinn decided to invent SmartCoat after quitting at JUGCO, the law would oblige Quinn to assign the SmartCoat patent to JUGCO if Quinn used anything learned at JUGCO to invent SmartCoat. JUGCO will also argue that Quinn is obliged to give JUGCO any idea for an innovation.

The **jury must decide** if Quinn did, or did not, use anything learned at JUGCO to invent SmartCoat or if Quinn was otherwise obliged to disclose the idea for a SmartCoat.

Plaintiff's (QUBCO's) council and witness goals:

1) In the opening statement tell the jury that direct testimony by Brook will show that:

a) The SmartCoat patent is valid and is infringed by JUGCO.

b) There is nothing in the SmartCoat patent which was learned at JUGCO.

c) Brook invented SmartCoat, not Quinn. Quinn only had a vague idea for a SmartCoat and the Illinois Employment Patent Act is about inventions (something claimed in a valid patent) and not about vague ideas.

2) In the direct testimony Brook, the plaintiff witness, convinces the jury of the points in 1) above.

3) In the cross examination discredit Jessiie and get Jessie to admit that neither the JUGCO employment contract nor the Illinois Employment Patent Act are about vague ideas nor even about innovations.

4) In the closing statement retell all the points in 1), 2), and 3) above and ask the jury to find for the plaintiff.

Defendant's (JUGCO's) council and witness goals:

1) In the <u>opening statement</u> tell the jury that direct testimony by Jessie will show that:

a) The SmartCoat patent is the property of JUGCO because all JUGCO employees are obliged to give ideas to JUGCO.

b) SmartCoat uses RFID which Quinn was exposed to at JUGCO.

2) In <u>direct testimony</u> Jessie convinces jury of the points in 1) above.

3) In <u>cross examination</u> discredit Brook and get Brook to admit that Quinn had an idea for an innovation.

4) In the <u>closing statement</u> retell all the points in 1), 2), and 3) above and ask the jury to find for the defendant.

[Remember that the attorneys and witnesses can not change the "facts" on the next page and in the depositions. The attorneys can not invent new "facts" either. Please review the sections of the general introduction which explain the scenarios and trials.]

[Also remember that attorneys can object if a witness is asked to testify about something not directly known by the witness and can object if a witness testifies about something not directly known by the witness.]

These "facts" are not disputed by plaintiff and defense and are entered as evidence: *[Y1 means year 1, Y2 year 2, Y3 year 3 – times are shorter than in reality for simplicity]*

Jan Y1: Quinn quits at JUGCO and then decides to invent SmartCoat to help friend Shelby keep from losing things. SmartCoat gives an alarm when RFID tagged things are not near the coat.

Feb Y1: Quinn and Brook incorporate as QUBCO. Brook works to perfect RFID used in SmartCoat.

Mar Y1: Shelby begins working at JUGCO.

Apr Y1: Jessie at JUGCO works on SmartyCoat.

Jan Y2: QUBCO makes and sells SmartCoat.

Feb Y2: JUGCO makes and sells SmartyCoat.

May Y2: QUBCO files patent application.

Jan Y3: QUBCO's patent granted

Feb Y3: QUBCO sues JUGCO for patent infringement.

The four depositions which follow are also "facts" which are entered as evidence.

[These "facts" are the raw material which the two attorney and witness teams use in opening statements, testimony, and closing statements.]

Deposition from the plaintiff's witness
Questioned by plaintiff's council
(Story for direct testimony)

[1] QUBCO Attorney: Please state your name and spell it for the record.
[2] Brook: Brook B R O O K.
[3] QUBCO Attorney: What is your occupation?
[4] Brook: I'm an inventor. People hire me to invent new products for them. I'm one of very few inventors succeeding that way.
[5] QUBCO Attorney: Who are you working for now?
[6] Brook: Quinn hired me to invent SmartCoat.
[7] QUBCO Attorney: Can you tell us what an invention is?
[8] Brook: An invention is a new product or process claimed in a valid patent.
[9] QUBCO Attorney: Do you have many patents?
[10] Brook: Yes, nearly a hundred.
[11] QUBCO Attorney: Who invented SmartCoat?
[12] Brook: I did. I invented everything in the claims.
[13] QUBCO Attorney: No one else contributed?
[14] Brook: No, not anything.
[15] QUBCO Attorney: Thank you, no further questions.

Deposition from the plaintiff's witness
Questioned by defendant's council
(Story for cross examination)

[1] JUGCO Attorney: An innovation is a new product or process which creats new value. Correct?
[2] Brook: Yes.
[3] JUGCO Attorney: Innovations are not always inventions. Correct?
[4] Brook: Yes.
[5] JUGCO Attorney: Quinn had the idea to invent SmartCoat. Correct?
[6] Brook: Yes.
[7] JUGCO Attorney: Since SmartCoat creates new value it is an innovation. Correct?
[8] Brook: Yes.
[9] JUGCO Attorney: SmartCoat is an innovation with and without the patent. Correct?
[10] Brook: I don't know.
[11] JUGCO Attorney: So, Quinn had an idea for an innovation which he did not disclose to his employer. Correct?
[12] Brook: I don't know.
[1] JUGCO Attorney: Thank you, no further questions.

Deposition from the defendant's witness

Questioned by defendant's council

(Story for direct testimony)

[1] JUGCO Attorney: Please state your name and spell it for the record.
[2] Jessie: Jessie. J E S S I E.
[3] JUGCO Attorney: What is your occupation?
[4] Jessie: I work for JUGCO.
[5] JUGCO Attorney: What is JUGCO?
[6] Jessie: We're a third-generation, family-owned manufacturer of clothing – been around for nearly a hundred years. When competition from abroad got tough we changed focus to higher quality and more innovative goods, like SmartyCoat.
[7] JUGCO Attorney: What is your job at JUGCO?
[8] Jessie: When we get an idea for an improved product, I work with all our other employees to invent enough improvements until we have a high quality innovation which does not infringe any patents and which will keep JUGCO going strong.
[9] JUGCO Attorney: Where do ideas for innovation come from?
[10] Jessie: From all over. Our employees are expected to suggest ideas for innovations. When we have an idea we want to explore our employees contribute to perfecting the idea.
[11] JUGCO Attorney: Are all employees expected to suggest ideas and contribute to perfecting ideas?
[12] Jessie: Yes, we all have a duty to to this.
[13] JUGCO Attorney: Does JUGCO use RFID?
[14] Jessie: Yes, we use RFID in our warehouses department where Quinn worked. We've had some nifty ideas from people in our warehouses department.

JUGCO Attorney: Thank you, no further questions.

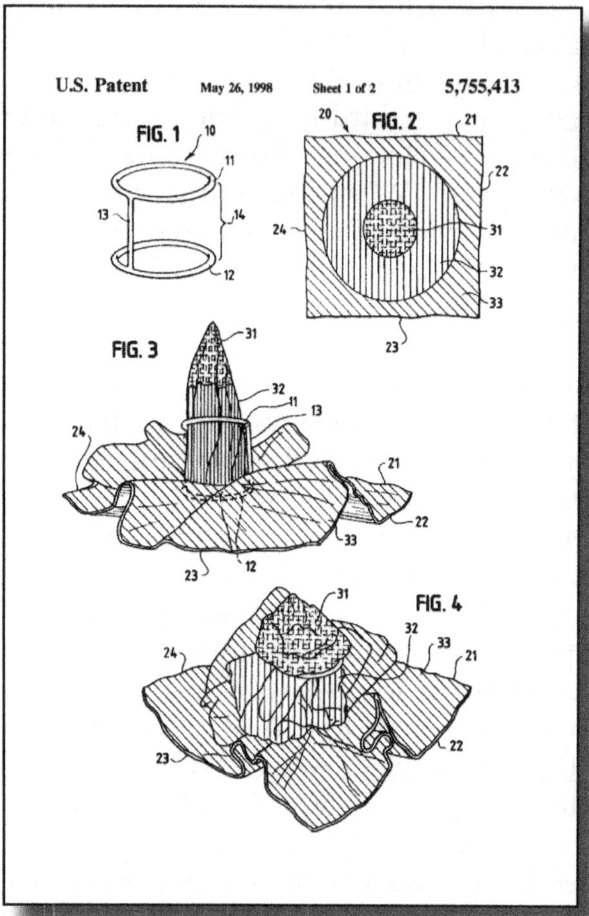

Deposition from the defendant's witness
Questioned by plaintiff's council
(Story for cross examination)

[1] QUBCO Attorney: There is nothing in the JUGCO employment contract about any duty to disclose ideas for innovations. Correct?
[2] Jessie: It's part of our culture. Everyone . . .
[3] QUBCO Attorney: Yes or no please. There is nothing about innovation in the employment contract. Correct? Yes or no.
[4] Jessie: Yes.
[5] QUBCO Attorney: The Illinois Employment Patent Act is about patents. Correct?
[6] Jessie: Yes.
[7] QUBCO Attorney: Innovations need not be inventions claimed in patents. Correct?
[8] Jessie: I don't know.
[9] QUBCO Attorney: So, Quinn was not obliged by JUGCO contract to disclose any ideas about innovations. Correct?
[10] Jessie: I don't know.
[11] QUBCO Attorney: And, Quinn did not make any invention so the Illinois Employment Patent Act does not apply. Correct?
[12] Jessie: I don't know.
[13] QUBCO Attorney: Thank you, no further questions.

Can you identify some innovation wrongs which led to this trial?

Can you suggest some ways those innovation wrongs could have been put right?

If you were head of QUBCO what might you do to improve QUBCO's innovation rights?

If you were head of JUGCO, what might you do to improve JUGCO's innovation rights?

SmartCoat Inequitable Conduct Trial

This scenario is a twist on the novelty and trade secret theft scenario which you might want to review. In that earlier scenario Jessie published everything in the SmartCoat patent application just over a year before QUBCO filed the SmartCoat patent application. Here, Jessie publishes just less than a year before QUBCO filed the SmartCoat patent application.

When Jessie publishes just over a year before the patent application is filed, then it counts as prior art causing the SmartCoat patent application to be rejected, even if the publication was based on trade secrets stolen from QUBCO. QUBCO had no recourse except to sue JUGCO for trade secret theft.

Under the new patent law passed late September 2011, Jessie's publication would not count as prior art and would not cause the SmartCoat patent application to be rejected, if Jessie's information about SmartCoat came directly or indirectly from QUBCO. We assume here that if the information was stolen, then it counts as coming directly, or indirectly from the inventor.

This presents QUBCO with a puzzle. QUBCO knows that Shelby stole their trade secrets and passed them on to Jessie, but can't prove it. If QUBCO tells the USPTO about the publication, then the SmartCoat patent application will be rejected, which would be unfair. QUBCO reasons that Jessie's publication is not relevant prior art and thus does not tell the USPTO about the publication.

Now, when QUBCO's SmartCoat patent issues and QUBCO sues JUGCO because SmartyCoat infringes the SmartCoat patent, JUGCO argues that the SmartCoat patent is not valid because QUBCO engaged in inequitable conduct with the USPTO by withholding information about Jessie's publication.

You can lose your patent right if you engage in inequitable conduct. If you fail to disclose to the USPTO information relevant to your invention, this could be inequitable conduct.

For example, it would be inequitable conduct with the USPTO if:

1) you do not tell the USPTO about a public disclosure of your invention which would lead to rejection of your claims, and also

2) you intended to deceive the USPTO to get a patent which is not deserved.

Inequitable conduct is especially slippery because it depends on intent, which is hard to determine. This is another heavy burden in patent practice.

This scenario provides an excellent opportunity for several observations about innovation rights:

1) Patent practitioners would like more clarity about inequitable conduct.

2) It is supremely important to understand all the rules and play by the rules, since any tiny mistake can lead to loss of valuable rights.

3) It will take years and much litigation to fully understand all the practical implications of the new patent law.

4) Uncertainty in practical implications of the law can be a barrier to innovation.

5) Understanding the law – including understanding uncertainties in the law – can lower barriers to innovation.

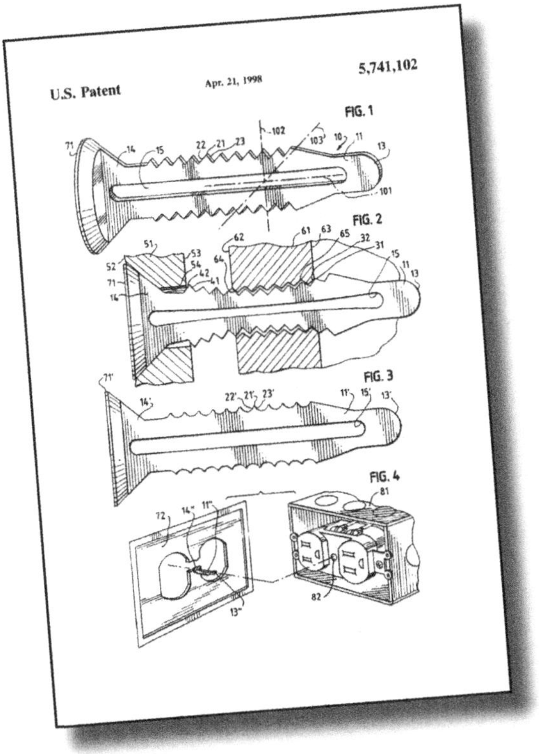

In our inequitable conduct case, **the jury must decide** if Quinn did or did not intend to deceive the USPTO in order to get a patent not deserved.

Plaintiff's (QUBCO's) council and witness goals:

1) In the opening statement tell the jury that direct testimony by Quinn will show that:

a) The SmartCoat patent is valid and JUGCO is infringing that patent.

b) There was no inequitable conduct because:

i) Jessie published information which was stolen from QUBCO

ii) information stolen from QUBCO does not count as prior art, so there was no need to disclose it.

iii) QUBCO did not obtain an undeserved patent.

2) In the direct testimony Quinn, the plaintiff's witness, convinces the jury of the points in 1) above.

3) In the cross examination discredit Jessie and get Jessie to admit that:

a) Jessie's publication contains the nonsense heater put in the QUBCO documents to catch thieves.

b) The information Jessie published came from Shelby.

4) In the closing statement retell all the points in 1), 2), and 3) above and ask the jury to find for the plaintiff.

Defendant's (JUGCO's) council and witness goals:

1) In the opening statement tell the jury that direct testimony by Jessie will show that:

a) The SmartCoat patent is not valid because the USPTO should have been told about the publication.

b) All of the information in Jessie's publication came from JUGCO employees and not from QUBCO.

2) In the direct testimony Jessie convinces jury of the points in 1) above.

3) In the cross examination discredit Quinn and get Quinn to admit that the information was knowingly withheld.

4) In the closing statement retell all the points in 1), 2), and 3) above and ask the jury to find for the defendant.

[Remember that the attorneys and witnesses can not change the "facts" on the next page and in the depositions. The attorneys can not invent new "facts" either. Please review the sections of the general introduction which explain the scenarios and trials.]

[Also remember that attorneys can object if a witness is asked to testify about something not directly known by the witness and can object if a witness testifies about something not directly known by the witness.]

These "facts" are not disputed by plaintiff and defense and are entered as evidence: *[Y1 means year 1, Y2 year 2, Y3 year 3 – times are shorter than in reality for simplicity]*

Jan Y1: Quinn decides to invent SmartCoat to help friend Shelby keep from losing things. SmartCoat gives an alarm when RFID tagged things are not near the coat.

Feb Y1: Quinn and Brook incorporate as QUBCO. Brook works to perfect RFID used in SmartCoat.

Mar Y1: Shelby begins working at JUGCO.

Jun Y1: Jessie at JUGCO works on SmartyCoat and publishes everything which will be claimed in QUBCO patent application in *Design News*.

Jan Y2: QUBCO makes and sells SmartCoat.

Feb Y2: JUGCO makes and sells SmartyCoat.

May Y2: QUBCO files patent application.

Jan Y3: QUBCO's patent issued.

Feb Y3: QUBCO sues JUGCO for infringment.

The four depositions which follow are also "facts" which are entered as evidence.

[These "facts" are the raw material which the two attorney and witness teams use in opening statements, testimony, and closing statements.]

Deposition from the plaintiff's witness
Questioned by plaintiff's council
(Story for direct testimony)

QUBCO Attorney: Please tell us your name and spell it for us.
Quinn: Quinn Q U I N N
QUBCO Attorney: What is your occupation?
Quinn: I am the head of a start-up company. Our first product is SmartCoat. We have other products in our pipeline.
QUBCO Attorney: Is SmartCoat patented?
Quinn: Yes it is and JUGCO is infringing the patent by selling SmartyCoats.
QUBCO Attorney: Did you engage in any inequitable conduct to get the patent?
Quinn: No. We disclosed everything which was relevant to patentability?
QUBCO Attorney: Why did you not disclose Jessie's publication?
Quinn: That publication was based on information which was stolen from me and thus is not relevant to patentability.
QUBCO Attorney: How do you know that the information was stolen from you?
Quinn: Brook is my Chief Innovation Officer and is an old pro and knows all the tricks. Brook put into our documents a nonsense part, a heater coil which was not needed and did nothing. That nonsense part is in Jessie's publication. This is proof that Jessie's publication was from information stolen from us.
QUBCO Attorney: Why didn't you sue JUGCO for trade secret theft?
Quinn: We needed the patent to make our investors comfortable.

<u>QUBCO Attorney:</u> Did withholding information about Jessie's publication deceive the USPTO?
<u>Quinn:</u> No. Telling them would have deceived the USPTO that our application was not novel.
<u>QUBCO Attorney:</u> Why not tell the USPTO and add that the publication was information stolen from you?
<u>Quinn:</u> There is no procedure to do that. We did not want to delay the patent because of our investors.
<u>QUBCO Attorney:</u> Thank you. No further questions.

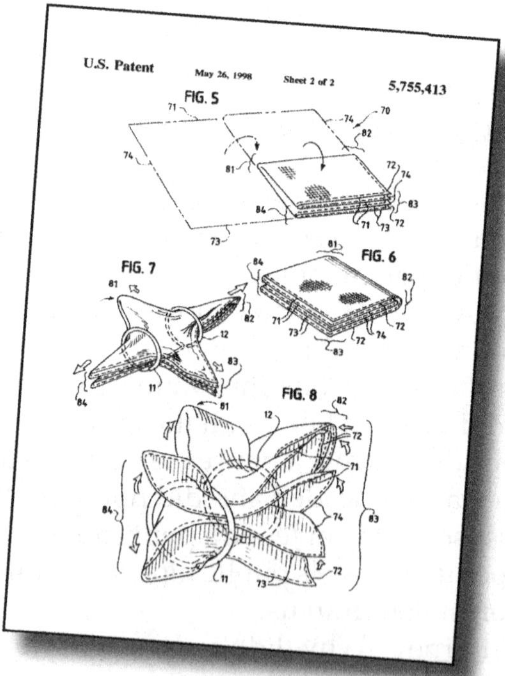

Deposition from the plaintiff's witness
Questioned by defendant's council
(Story for <u>cross examination</u>)

<u>JUGCO Attorney:</u> You knew about Jessie's publication when you filed the SmartCoat patent application. Correct?
<u>Quinn:</u> Yes.
<u>JUGCO Attorney:</u> You did not disclose Jessie's publication to the USPTO. Correct?
<u>Quinn:</u> Yes.
<u>JUGCO Attorney:</u> If you had disclosed Jessie's publication the USPTO would have rejected your application. Correct?
<u>Quinn:</u> I don't know.
<u>JUGCO Attorney:</u> You intended to withhold information which would have led to rejection of the SmartCoat patent application. Correct?
<u>Quinn:</u> I did not intend to get an patent we did not deserve.
<u>JUGCO Attorney:</u> Yes or no please.
<u>Quinn:</u> That is not correct because . . .
<u>JUGCO Attorney:</u> Thank you. No further questions.

Deposition from the defendant's witness
Questioned by defendant's council
(Story for direct testimony)

JUGCO Attorney: Please tell us your name and spell it for us.
Jessie: Jessie J E S S I E
JUGCO Attorney: What is your occupation?
Jessie: I work for JUGCO.
JUGCO Attorney: What is JUGCO?
Jessie: We were a third-generation, family-owned manufacturer of clothing – been around for nearly a hundred years. When competition from abroad got tough we changed focus to higher quality and more innovative goods, like SmartyCoat.
JUGCO Attorney: What is your job at JUGCO?
Jessie: When we get an idea for an improved product, I work with all our other employees to invent enough improvements until we have a high quality innovation which does not infringe any patents and which will keep JUGCO going strong.
JUGCO Attorney: Is that how your publication came about?
Jessie: Yes, I got lots of ideas from all of our employees and I put them together to invent SmartyCoat and published a detailed description in *Design News*.
JUGCO Attorney: Did you benefit from any information stolen from Quinn?
Jessie: I only got information from our employees.
JUGCO Attorney: So, no information in your publication came directly or indirectly from Quinn?
Jessie: I only got information from JUGCO employees.
JUGCO Attorney: Thank you. No further questions.

Deposition from the defendant's witness
Questioned by plaintiff's council
(Story for cross examination)

JUGCO Attorney: You put together ideas from JUGCO employees for your publication. Correct?
Jessie: Yes.
JUGCO Attorney: You do not know where those employees got the information. Correct?
Jessie: JUGCO employees are all honest . . .
JUGCO Attorney: Yes or no please.
Jessie: No.
JUGCO Attorney: Shelby was a JUGCO employee when you were gathering information. Correct?
Jessie: Yes.
JUGCO Attorney: Shelby might have fed you information stolen from Quinn. Correct?
Jessie: I don't know.
JUGCO Attorney: It is possible. Correct?
Jessie: I guess so.
JUGCO Attorney: There is a heater coil in the base module of SmartyCoat in your publication. Correct?
Jessie: Yes.
JUGCO Attorney: It is not needed though. Correct?
Jessie: I don't remember.
JUGCO Attorney: You got it from Shelby. Correct?
Jessie: I don't remember.
JUGCO Attorney: Thank you. No further questions.

Can you identify some innovation wrongs which led to this trial?

Can you suggest some ways those innovation wrongs could have been put right?

If you were head of QUBCO what might you do to improve QUBCO's innovation rights?

If you were head of JUGCO, what might you do to improve JUGCO's innovation rights?

Part Two

Part Two Introduction

Because SmartCoat was not enough to grow QUBCO, Quinn and Brook focused attention on new ideas for products. In one of their brainstorming sessions they generated ideas for SmartFoot, an insert for shoes which would harvest energy as people walked during the day to charge a battery. At night the battery could charge a cell phone and similar devices.

Brook perfected the idea using piezoelectric elements to convert energy of walking into electricity to charge the battery. Soon SmartFoot was on sale.

QUBCO did not file a patent application on SmartFoot because they were short of money and because Quinn decided that patents were just licenses to sue.

Brook argued that they could be the ones sued if they did not file a patent application. Quinn overruled that argument. Brook right – Quinn wrong.

JUGCO bought a SmartFoot for testing. As usual all JUGCO workers took part in the testing. They were highly critical because they thought that SmartFoot could be greatly improved. The JUGCO innovation process went to full speed.

Because JUGCO had a very diverse workforce, because those diverse workers cooperated excellently, and because the competition for generating ideas was stimulating and fun, the JUGCO innovation process worked extremely well and soon generated major improvements.

The much improved product was named SmartyFoot. Improvements were made in the day-time energy harvesting, in the connections for night-time charging of devices, in the size and shape so the insert fit better in more shoes, and to ease manufacturing lowering the cost.

Rowan was the third generation head at JUGCO, after the death of Emerson, the second generation head. With SmartyFoot Rowan heeded Emerson's advice to pay attention to JUGCO's innovation rights. Thus JUGCO filed a patent application covering all the improvements in SmartyFoot.

JUGCO made and sold SmartyFoot which did sell like hot cakes. Who would want a SmartFoot when the much better SmartyFoot was available for less cost?

QUBCO could do nothing to stop JUGCO because QUBCO had not patented SmartFoot. Quinn decided that QUBCO should improve SmartFoot by adding all the changes made by JUGCO.

Brook pointed out that they could be sued by JUGCO. Quinn argued that they would show that the JUGCO patent was invalid because SmartFoot was on sale before JUGCO filed a patent application.

JUGCO sued QUBCO for infringing the SmartyCoat patent.

That trial could have at least three outcomes:

QUBCO could prevail with the argument that the SmartyCoat patent is invalid because SmartCoat anticipated claims in the SmartyFoot patent. Then there would be no patent protection for anyone.

JUGCO could prevail.

JUGCO and QUBCO could settle with JUGCO licensing to QUBCO rights to make and sell SmartyCoat. And, as part of the settlement JUGCO and QUBCO could agree to a joint venture to develop new products. JUGCO would get access to Brook's extraordinary inventive abilities. QUBCO would get access to JUGCO's extraordinary group innovation process.

Each of these outcomes could lead to more innovation wrongs and new innovation rights struggles if the parties are not careful. Others, such as SHELBYCO (remember Shelby), can also make mischief.

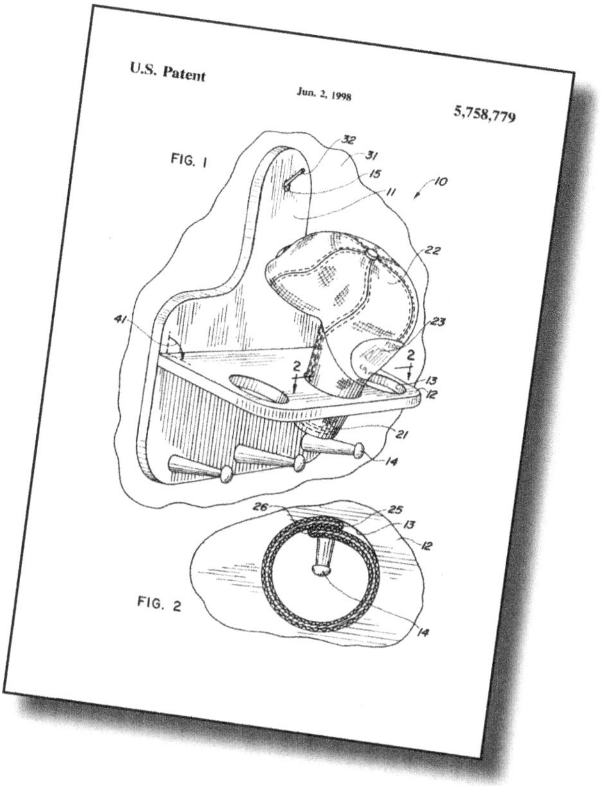

SmartyFoot Novelty Trial

QUBCO created, made, and sold SmartFoot which is a shoe insert using piezoelectric elements to harvest energy of walking and running to charge a battery. That battery is then used at night to charge smart phones and like devices.

QUBCO did not file a patent application for SmartFoot. JUGCO copied SmartFoot and made improvements to make and sell SmartyFoot which worked better, lasted longer, fit more shoes, and cost much less.

QUBCO copied and made and sold SmartyFoot. JUGCO did file a patent application covering the SmartyFoot improvements. When the SmartyFoot patent grant issued, JUGCO sued QUBCO for infringing the SmartyFoot patent.

Your patent is infringed when someone sells a product which has all the parts in a claim in your patent. You must sue the infringer to make them stop and to get damages. The infringer can argue that their product is not equivalent to the patented product. And, the infringer can argue that the patent is not valid.

QUBCO's SmartyFoot does have all the parts in JUGCO's SmartyFoot patent claims. QUBCO will argue in the trial that the SmartyFoot patent is not valid because all the improvements JUGCO made to QUBCO's SmartFoot are not novel, just old stuff applied to SmartFoot.

In this case **the jury must decide** if the improvements made by QUBCO do, or do not, add up to something new.

Plaintiff's (JUGCO's) council and witness goals:

1) In the opening statement tell the jury that the direct testimony by Jessie will show that:

   a) The SmartyCoat patent is valid and is infringed by QUBCO.
   b) There was no fault in the examination of the SmartyFoot patent application.

2) In the direct testimony Jessie convinces the jury of the points in 1) above.

3) In the cross examination discredit Brook and get Brook to admit that:

   a) The direct testimony by Brook is not trustworthy because there is no evidence supporting what Brook says.
   b) Brook is not trustworthy because Brook has a strong motivation to testify against the validity of the patent.

4) In the closing statement retell all the points in 1), 2), and 3) above and ask the jury to find for the plaintiff.

Defendant's (QUBCO's) council and witness goals:

1) In the opening statement tell the jury that direct testimony by Brook will show that the SmartyFoot patent is not valid because the examiner did not search shoe making art and thus did not find the art which would have made the claimed invention not novel.

2) In the direct testimony Brook convinces the jury of the points in 1) above.

3) In cross examination discredit Jessie and get Jessie to admit that

a) The direct testimony by Jessie is not trustworthy because it is not about shoe making art.

b) Jessie is not trustworthy because the patent application hid the fact that SmartyFoot is a shoe insert.

4) In the closing statement retell all the points in 1), 2), and 3) above and ask the jury to find for the defendant.

[Remember that the attorneys and witnesses can not change the "facts" on the next page and in the depositions. The attorneys can not invent new "facts" either. Please review the sections of the general introduction which explain the scenarios and trials.]

[Also remember that attorneys can object if a witness is asked to testify about something not directly known by the witness and can object if a witness testifies about something not directly known by the witness.]

These "facts" are not disputed by plaintiff and defense and are entered as evidence:

Jan Y1: QUBCO makes and sells SmartFoot with no patent protection. *[Y1 = year one, Y2 = year 2]*

Feb Y1: JUGCO files patent application and makes and sells SmartyFoot.

Mar Y1: QUBCO makes and sells SmartyFoot

Jan Y2: SmartyFoot patent granted. JUGCO sues QUBCO for infringement.

Feb Y2: **Novelty Trial** with QUBCO in defense arguing SmartyFoot is not novel.

[These events are just the beginning of a longer story with lots of twists and turns and tangles.]

The four depositions which follow are also "facts" which are entered as evidence.

[These "facts" are the raw material which the two attorney and witness teams use in opening statements, testimony, and closing statements.]

Deposition from the plaintiff's witness
Questioned by plaintiff's council
(Story for direct testimony)

[1] JUGCO Attorney: Please tell us your name and spell it for the record.
[2] Jessie: Jessie J E S S I E
[3] JUGCO Attorney: What is your occupation?
[4] Jessie: I work for JUGCO. I coordinate our innovation process.
[5] JUGCO Attorney: In that capacity, did you oversee the patenting of SmartyFoot?
[6] Jessie: Yes, I did. I worked with our patent agent to help write the patent application. And, I worked with our patent agent to respond to the office action by the USPTO examiner.
[7] JUGCO Attorney: Do you oversee many patentings?
[8] Jessie: Yes, at least a dozen. Very familiar with the process now.
[9] JUGCO Attorney: Tell us how it works?
[10] Jessie: You file a patent application. First it is checked to see that all the formalities are OK. Then it is assigned to a patent examiner working in the relevant field of art. And, you wait as long as three years for the examiner's first office action.
[11] JUGCO Attorney: What does the patent examiner do?
[12] Jessie: The examiner searches prior art relevant to what is claimed in the patent application. If the examiner finds that a claim is anticipated by prior art, then the examiner rejects the claim. If the examiner sees that a claim is just an obvious change from the prior art, then the examiner rejects the claim.

[13] JUGCO Attorney: What happened with the first office action for SmartyFoot?
[14] Jessie: First, I was impressed at how thoroughly the examiner searched the prior art. The examiner thought that our claims were anticipated by this prior art and rejected them.
[15] JUGCO Attorney: What did you do?
[16] Jessie: We explained to the examiner that the parts of SmartyFoot which the examiner thought were anticipated actually worked very differently together than in the prior art. So, the examiner dropped the novelty rejection.
[17] JUGCO Attorney: Anything else?
[18] Jessie: The examiner also thought that those parts of SmartyFoot were just obvious changes in the prior art and rejected the claims on that basis.
[19] JUGCO Attorney: What did you do?
[20] Jessie: We explained to the examiner that it was an unexpected discovery that the parts in question worked together as they did, so the examiner dropped this rejection also.
[21] JUGCO Attorney: Is this typical?
[22] Jessie: In my experience yes. The examiner did a thorough job to help us get the best patent we are entitled to.
[23] JUGCO Attorney: So JUGCO has a patent grant covering SmartyFoot.
[24] Jessie: Yes, and according to the law the patent is presumed valid.
[25] JUGCO Attorney: From your experience does the patent examiner's work justify that presumption of validity?
[26] Jessie: It sure does. We have a patent and it is a valid patent.
[27] JUGCO Attorney: Thank you. No further questions.

Deposition from the plaintiff's witness
Questioned by defendant's council
(Story for cross examination)

[1] QUBCO Attorney: The examiner searched only art relevant to energy harvesting. Correct? Or not correct?
[2] Jessie: Correct.
[3] QUBCO Attorney: The examiner did not search any shoe-making art. Correct? Or not correct?
[4] Jessie: Correct.
[5] QUBCO Attorney: In the claims of the SmartyFoot patent application you did not mention that the product would be a shoe insert. Correct? Or not correct?
[6] Jessie: Yes.
[7] QUBCO Attorney: In the written description of the SmartyFoot patent application you did not mention that the product would be a shoe insert. Correct? Or not correct?
[8] Jessie: Correct.
[9] QUBCO Attorney: If you had mentioned that the product would be a shoe insert, then the examiner would also have searched shoe-making art. Correct? Or not correct?
[10] Jessie: I don't know.
[11] QUBCO Attorney: Thank you. No further questions.

Deposition from the defendant's witness
Questioned by defendant's council
(Story for direct testimony)

[1] QUBCO Attorney: Please state your name and spell it for the record.
[2] Brook: Brook, B R O O K.
[3] QUBCO Attorney: What is your occupation?
[4] Brook: I am a co-owner of QUBCO. I invent our products.
[5] QUBCO Attorney: Did you invent SmartFoot?
[6] Brook: Yes, I did. It is the second QUBCO product I invented.
[7] QUBCO Attorney: Is there any part of SmartyFoot which surprised you?
[8] Brook: No, it's all old stuff which I was going to add to SmartFoot.
[9] QUBCO Attorney: Why didn't you add the old stuff to SmartFoot before you put SmartFoot on the market?
[10] Brook: Quinn, the other owner of QUBCO was in a hurry to get the product on the market. I told him that somebody else would make the improvements, but Quinn didn't listen.
[11] QUBCO Attorney: How do you know that SmartyFoot is just old stuff added to SmartFoot?
[12] Brook: I did some invention for a shoe manufacturer and learned much about that industry. The so-called improvements JUGCO made in SmartFoot are all well known in the shoe-making art.
[13] QUBCO Attorney: Where should a patent examiner look to find this prior art?
[14] Brook: The examiner should look at shoe-making art.

[15] QUBCO Attorney: How did you learn this shoe-making art?
[16] Brook: When I was doing some inventing for a shoe manufacturer I went to the shop floor and got to know the workers. Watched what they were doing. Asked questions. Had beers with them after quitting time.
[17] QUBCO Attorney: Why didn't QUBCO patent SmartFoot?
[18] Brook: Quinn said that patents are just a right to sue and just trouble. I told him that we could be the one sued which would be even more trouble, but Quinn didn't listen.
[19] QUBCO Attorney: Did Quinn say that if QUBCO was sued for infringing a patent related to SmartFoot, that QUBCO could argue that the patent was not valid because SmartFoot was first.
[20] Brook: Yes, Quinn said that. I thought it would be better to patent SmartFoot to avoid trouble, but I didn't fight about it because I agreed that any patent related to SmartFoot would not be valid because SmartFoot was first and any improvements would be old stuff.
[21] QUBCO Attorney: Thank you. No further questions.

Deposition from the defendant's witness
Questioned by plaintiff's council
(Story for cross examination)

[1] JUGCO Attorney: You have much experience with patenting. Correct? Or not correct?
[2] Brook: Correct.
[3] JUGCO Attorney: It is true that even though a device is patented, improvements on that device can be patented. Correct? Or not correct?
[4] Brook: Only if the improvements are novel.
[5] JUGCO Attorney: Correct or not correct please.
[6] Brook: Correct.
[7] JUGCO Attorney: You say that the improvements to SmartCoat which JUGCO patented are all old stuff, but we have no other evidence to support what you say. Correct? Or not correct?
[8] Brook: Correct.
[9] JUGCO Attorney: So, we only have your word on this. Correct? Or not correct?
[10] Brook: My word is good as gold.
[11] JUGCO Attorney: Yor are a co-owner of QUBCO. Correct? Or not correct?
[12] Brook: Correct.
[13] JUGCO Attorney: If JUGCO's patent is valid, then QUBCO is stuck with SmartCoat which costs more than SmartyCoat and does not work as well so your income goes way down. Correct? Or not correct?
[14] Brook: I don't know.
[15] JUGCO Attorney: Thank you. No further questions.

Can you identify some innovation wrongs which led to this trial?

Can you suggest some ways those innovation wrongs could have been put right?

If you were head of QUBCO what might you do to improve QUBCO's innovation rights?

If you were head of JUGCO, what might you do to improve JUGCO's innovation rights?

SmartyPedo Obvious Trial

QUBCO invented SmartFoot but did not patent. JUGCO improved SmartFoot and patented SmartyFoot. QUBCO copied SmartyFoot so JUGCO sued QUBCO.

QUBCO argued that SmartyFoot was not novel because of SmartFoot and won. There was then no patent on SmartyFoot.

QUBCO added a pedometer to SmartyFoot and patented SmartyPedo. JUGCO copied SmartyPedo so QUBCO sues JUGCO for infringement.

Your patent is infringed when someone sells a product which has all the parts in a claim in your patent. You must sue the infringer to make them stop and to get damages. The infringer can argue that their product is not equivalent to the patented product. And, the infringer can argue that the patent is not valid.

JUGCO will argue that the SmartyPedo patent is not valid because it just adds obvious changes to existing art.

So, in this case the **jury must decide** if the claims in the SmartyPedo patent would be, or would not be, obvious to a person having ordinary skill in the relevant art at the time the patent application was filed.

[Obviousness determinations are incredibly slippery and are a great burden on examiners and the whole patent system. How the jury decides in this trial will depend very much on closing statements.]

Plaintiff's (QUBCO's) council and witness goals:

1) In the opening statement tell the jury that the direct testimony by Brook will show that:

a) The SmartyPedo patent is valid and is infringed by JUGCO.
b) The SmartyPedo patent claims a new and unexpected result and thus is not obvious.

2) In the direct testimony Brook convinces the jury of the points in 1) above.

3) In the cross examination discredit Casey and get Casey to admit that:

a) The direct testimony by Casey is not trustworthy because, even though there are only so many ways to combine two signals, the data in the two signals can produce unexpected new results when combined;
b) Casey is not trustworthy because Casey recently applied for a position at the law firm which does work for JUGCO;

4) In the closing statement retell all the points in 1), 2), and 3) above and ask the jury to find for the plaintiff.

Defendant's (JUGCO's) council and witness goals:

1) In the opening statement tell the jury that direct testimony by Casey will show that the SmartyFoot patent is not valid because there are only so many ways to combine two signals

2) In the direct testimony Casey convinces jury of the points in 1) above.

3) In cross examination discredit Brook and get Brook to admit that:

a) The direct testimony by Brook is not trustworthy because Brook has no direct experience nor training in patent examination.

b) Brook is not trustworthy because Brook has a strong interest in the outcome;

4) In the closing statement retell all the points in 1), 2), and 3) above and ask the jury to find for the defendant.

[Remember that the attorneys and witnesses can not change the "facts" on the next page and in the depositions. The attorneys can not invent new "facts" either. Please review the sections of the general introduction which explain the scenarios and trials.]

[Also remember that attorneys can object if a witness is asked to testify about something not directly known by the witness and can object if a witness testifies about something not directly known by the witness.]

These "facts" are not disputed by plaintiff and defense and are entered as evidence:

Jan Y1: QUBCO makes and sells SmartFoot no patent protection. *[Y1 = year one Y2 = year two Y3 = year three]*

Feb Y1: JUGCO files patent application and makes and sells SmartyFoot.

Mar Y1: QUBCO makes and sells SmartyFoot.

Jan Y2: SmartyFoot patent granted. JUGCO sues QUBCO for infringement.

Feb Y2: Novelty trial with QUBCO in defense arguing SmartyFoot is not novel. QUBCO wins the novelty trial. Thus there is no valid patent for SmartyFoot.

Mar Y2: QUBCO adds pedometer to SmartyFoot to get SmartyPedo, applies for a patent, and makes and sells SmartyPedo.

Jul Y2: JUGCO makes and sells SmartyPedo.

Feb Y3: SmartyPedo patent granted to QUBCO. QUBCO sues JUGCO for infringing SmartyPedo patent.

Mar Y3: **Trial** with JUGCO arguing that SmartyPedo is an obvious change to SmartyFoot.

The four depositions which follow are also "facts" which are entered as evidence.

Deposition from the plaintiff's witness
Questioned by plaintiff's council
(Story for direct testimony)

[1] QUBCO Attorney: Please tell us your namd and spell it for the record.
[2] Brook: Brook B R O O K.
[3] QUBCO Attorney: What is your occupation?
[4] Brook: I am part owner of QUBCO. I invent our advanced technology products.
[5] QUBCO Attorney: Do you have much experience with patenting?
[6] Brook: Yes, my first patent was issued about fifty years ago. I average just over one patent per year since then. Worked with many patent agents and patent attorneys. Many changes in the law over that time.
[7] QUBCO Attorney: How do you present inventions to overcome the current USPTO obviousness guidelines?
[8] Brook: Often the invention solves a problem not recognized before. We emphasize that. We point out that there is a new result. That the parts of the invention work together in an unexpected way. That discoveries were made along the way to the invention.
[9] QUBCO Attorney: How did this apply to SmartyPedo?
[10] Brook: Using piezoelectric elements in each of two shoes made for new opportunities for gathering information and presented new problems for making that information useful. Looking at the combined signals from the two shoe inserts I saw patterns I had not expected. It turned out that those patterns could be used to

analyze running, or walking, styles and suggest changes which would cause less wear and tear on bodies.

[11] QUBCO Attorney: Was this all presented in the patent application?

[12] Brook: Of course. Unexpected problem discovered. Unexpected new results discovered.

[13] QUBCO Attorney: Did the USPTO examiner reject any claims as obvious?

[14] Brook: No.

[15] QUBCO Attorney: From your long experience patenting what is your understanding of the status of an issued patent?

[16] Brook: An issued patent is presumed valid. Recent changes in the law have strengthened the presumption of validity.

[17] QUBCO Attorney: Again based on your fifty years of experience patenting, what is your understanding of the requirements for finding a patent not valid because of obviousness.

[18] Brook: My understanding is that this is an uphill battle. That the other side has a heavy burden to show how the USPTO erred.

[19] QUBCO Attorney: Thank you. No further questions.

Deposition from the plaintiff's witness

Questioned by defendant's council

(Story for cross examination)

[1] JUGCO Attorney: You have no experience as a patent examiner. Correct? Or, not correct?
[2] Brook: Correct.
[3] JUGCO Attorney: You have not been trained nor supervised as a patent examiner. Correct? Or, not correct?
[4] Brook: Correct.
[5] JUGCO Attorney: You are not a patent attorney. Correct? Or, not correct?
[6] Brook: Correct.
[7] JUGCO Attorney: You never studied the USPTO Manual of Patent Examining Procedures. Correct? Or, not correct?
[8] Brook: Correct.
[9] JUGCO Attorney: You can not prove that SmartyPedo is not obvious. Correct? Or, not correct?
[10] Brook: No one can prove a negative.
[11] JUGCO Attorney: Thank you. No further questions.

Deposition from the defendant's witness
Questioned by defendant's council
(Story for direct testimony)

[1] JUGCO Attorney: Tell us your name and spell it for the record.
[2] Casey: Casey C A S E Y.
[3] JUGCO Attorney: What is your occupation?
[4] Casey: I am a patent attorney with a solo practice.
[5] JUGCO Attorney: What did you do before that?
[6] Casey: I was a patent examiner at the USPTO for five years.
[7] JUGCO Attorney: Tell us how you examined patent applications for obviousness when you were an examiner.
[8] Casey: First we look at the claims. Then we search the prior art. When we see that the claims are anticipated in the prior art, then we reject the claim as not novel. When we see that the claims are obvious changes in prior art, then we reject the claim as obvious.
[9] JUGCO Attorney: What tests did you use to determine obvious?
[10] Casey: If there are only so many ways to solve a problem and if those ways are expected to work, then picking one of those ways is obvious.
[11] JUGCO Attorney: How does this apply to SmartyPedo?
[12] Casey: You have signals from two SmartyFoot shoe inserts. There are only so many ways to combine those signals and they all work somehow. So, it's obvious.
[13] JUGCO Attorney: No new result?

[14] Casey: Can't be.
[15] JUGCO Attorney: No discovery?
[16] Casey: Can't be.
[17] JUGCO Attorney: Nothing unexpected?
[18] Casey: Can't be.
[19] JUGCO Attorney: Thank you. No further questions.

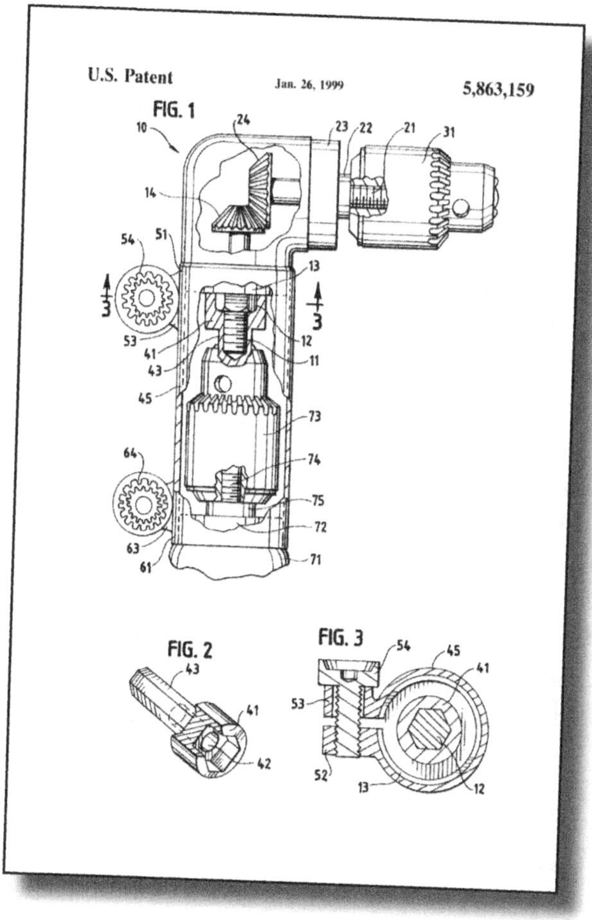

Deposition from the defendant's witness
Questioned by plaintiff's council
(Story for cross examination)

[1] QUBCO Attorney: Patent examiners are very hard pressed for time to examine patents. Correct? Or, not correct?
[2] Casey: Correct.
[3] QUBCO Attorney: Which means that when examiners see a reason to reject they do so without further study. Correct? Or, not correct?
[4] Casey: They do the best they can.
[5] QUBCO Attorney: Correct? Or, not correct? Please.
[6] Casey: Correct.
[7] QUBCO Attorney: You recently applied for a job with the law firm which does lots of work for JUGCO. Correct? Or, not correct?
[8] Casey: That has nothing to do with anything.
[9] QUBCO Attorney: Correct? Or, not correct? Please.
[10] Casey: Correct.
[11] QUBCO Attorney: You can not present any evidence that the patterns in the combinations of the two signals described and claimed in the SmartyPedo patent were expected by persons skilled in the art. Correct? Or, not correct?
[12] Casey: Correct?
[13] QUBCO Attorney: You just assumed that they were expected because there are only so many ways to combine signals. Correct? Or, not correct?
[14] Casey: Correct.

[15] QUBCO Attorney: Even though there are only so many ways to combine signals there can be unexpected results. Correct? Or, not correct?
[16] Casey: I don't know.
[17] QUBCO Attorney: Thank you. No further questions.

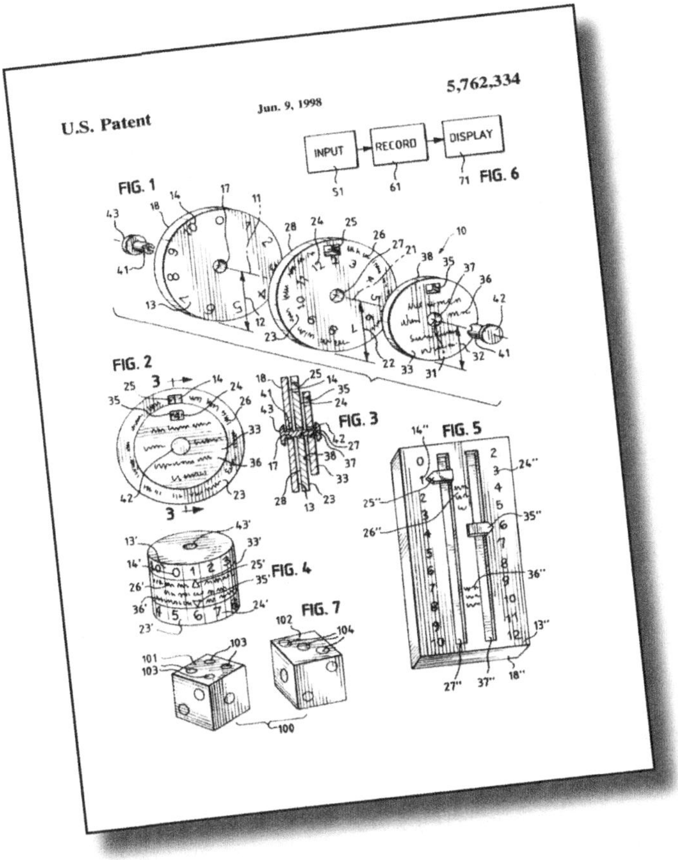

Can you identify some innovation wrongs which led to this trial?

Can you suggest some ways those innovation wrongs could have been put right?

If you were head of QUBCO what might you do to improve QUBCO's innovation rights?

If you were head of JUGCO, what might you do to improve JUGCO's innovation rights?

SmartyPedo Infringe Trial

QUBCO invented SmartFoot but did not patent. JUGCO improved SmartFoot and patented SmartyFoot.

QUBCO copied SmartyFoot so JUGCO sued QUBCO. QUBCO argued that SmartyFoot was not novel. JUGCO prevailed. The JUGCO SmartyFoot patent was valid. QUBCO ceased making and selling SmartyFoot.

QUBCO added a pedometer to SmartyFoot and began making and selling SmartyPedo, pat pend.

JUGCO sued QUBCO for infringing JUGCO's SmartyFoot patent.

Your patent is infringed when a competitor sells a product which has all the elements in any of your claims. You must sue the competitor to stop the competitor and to collect damages. If the competitor has all the elements in any of your claims and other elements as well, then that competitor infringes.

In this case the **jury must decide** if SmartyPedo does – or does not – have all the elements in a SmartyFoot claim. QUBCO will argue that the tiny "battery" in SmartyPedo is not the same as the high capacity "battery" in SmartyFoot.

Plaintiff's (JUGCO's) council and witness goals:

1) In the opening statement tell the jury that direct testimony by Jaime will show that:

a) The SmartPedo patent is valid and is infringed by QUBCO;

b) The SmartyPedo battery performs substantially the same *function* as SmartyFoot battery in substantially the same *way* to yield substantially the same *result*.

2) In the direct testimony Jaime, the plaintiff witness, convinces the jury of the points in 1) above.

3) In the cross examination discredit Harley and get Harley to admit that:

a) The direct testimony by Harley is not trustworthy because Harley confused how the batteries functioned with the function they preform;

b) Harley is not trustworthy because Harley is an admitted liar;

4) In the closing statement retell all the points in 1), 2), and 3) above and ask the jury to find for the plaintiff.

Defendant's (QUBCO's) council and witness goals:

1) In the opening statement tell the jury that direct testimony by Harley will show that SmartyPedo does not infringe the SmartyFoot patent because the tiny primary battery in SmartyPedo functions differently from the large secondary battery in SmartyFoot.

2) In direct testimony Harley convinces jury of the points in 1) above.

3) In cross examination discredit Jaime and get Jaime to admit that:

a) The direct testimony by Jaime is not trustworthy because Jamie does not know how batteries work.
b) The Jaime is not trustworthy because Jamie has a scholarship to law school paid for by JUGCO.

4) In the closing statement retell all the points in 1), 2), and 3) above and ask the jury to find for the defendant.

[Remember that the attorneys and witnesses can not change the "facts" on the next page and in the depositions. The attorneys can not invent new "facts" either. Please review the sections of the general introduction which explain the scenarios and trials.]

[Also remember that attorneys can object if a witness is asked to testify about something not directly known by the witness and can object if a witness testifies about something not directly known by the witness.]

These "facts" are not disputed by plaintiff and defense and are entered as evidence:

Jan Y1: QUBCO makes and sells SmartFoot no patent protection. *[Y1 = year one Y2 = year two Y3 = year three]*

Feb Y1: JUGCO files patent application and makes and sells SmartyFoot.

Mar Y1: QUBCO makes and sells SmartyFoot.

Jan Y2: SmartyFoot patent granted. JUGCO sues QUBCO for infringement.

Feb Y2: Novelty trial with QUBCO in defense arguing SmartyFoot is not novel. JUGCO prevails. The JUGCO SmartyFoot patent is valid.

Mar Y2: QUBCO adds pedometer to SmartyFoot to get SmartyPedo and makes and sells SmartyPedo. JUGCO sues QUBCO for infringing the SmartyFoot patent

Apr Y2: JUGCO makes and sells SmartyPedo.

Feb Y3: **Infringe trial** with QUBCO arguing that SmartyPedo does not infringe SmartyFoot.

The four depositions which follow are also "facts" which are entered as evidence.

Deposition from the plaintiff's witness
Questioned by plaintiff's council
(Story for <u>direct testimony</u>)

<u>[1] JUGCO Attorney:</u> Tell us your name and spell it for the record.
[2] Jamie J A M I E.
<u>[3] JUGCO Attorney:</u> What is your occupation?
<u>[4] Jamie:</u> I am a full-time student in law school preparing to be a patent attorney. I'm finished next month.
<u>[5] JUGCO Attorney:</u> So, can you tell us about the patent law and about how infringement is determined?
<u>[6] Jamie:</u> A product infringes a patent if the product has all the elements in a claim of the patent. The product can have more elements and it will still infringe when it has all the elements in the patent.
<u>[7] JUGCO Attorney:</u> QUBCO admits that SmartyPedo has all the elements of SmatyFoot except that they argue that the battery in SmartyPedo is not the same as the battery in SmartyFoot. What do you say to that?
<u>[8] Jamie:</u> The claim in the SmartyFoot patent has a battery and the claim does not say what kind of battery.
<u>[9] JUGCO Attorney:</u> Does it make any difference in the law that the battery in SmartyPedo is smaller than the battery in SmartyPedo?
<u>[10] Jamie:</u> No.
<u>[11] JUGCO Attorney:</u> So, you are saying that SmartyPedo has all the elements in the SmartyFoot claim and more and thus it infringes the SmartyFoot patent. correct?

[12] Jamie: Yes.
[13] JUGCO Attorney: Thank you. No further questions.

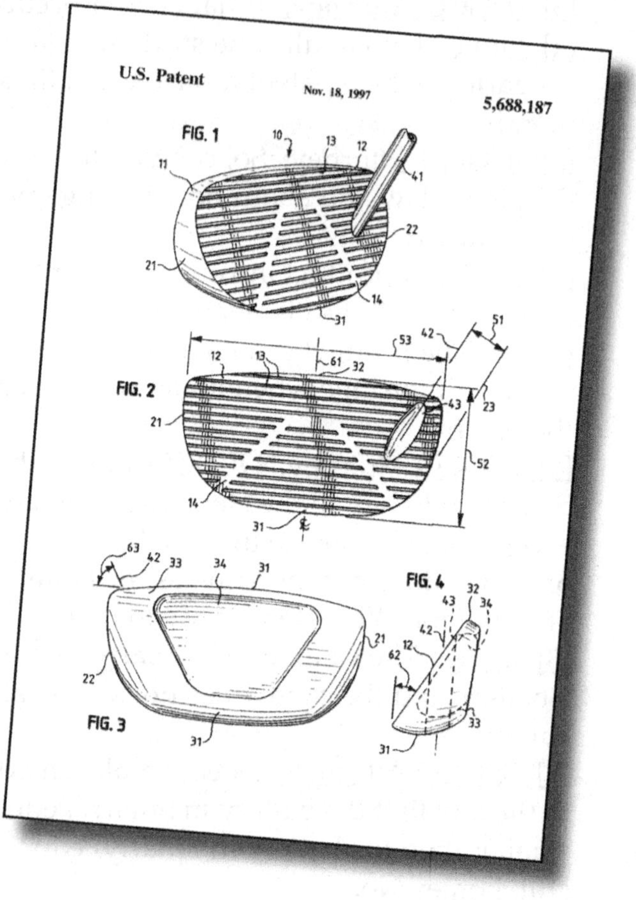

Deposition from the plaintiff's witness
Questioned by defendant's council
(Story for cross examination)

[1] QUBCO Attorney: Your undergraduate degree is in neuroscience. Correct? Or, not correct?
[2] Jamie: Correct.
[3] QUBCO Attorney: You have no expertise in battery art. Correct? Or, not correct?
[4] Jamie: Correct.
[5] QUBCO Attorney: The test for an element in SmartyPedo to be equivalent to an element in a SmartyFoot claim is that the element must preform the same function in the same way to yield the same result. Correct? Or, not correct?
[6] Jamie: Correct.
[7] QUBCO Attorney: The function of the SmartyFoot battery is to store charge in order to charge a cell phone. Correct? Or, not correct?
[8] Jamie: Correct.
[9] QUBCO Attorney: The battery in SmartyPedo can not have this function. Correct? Or, not correct?
[10] Jamie: I don't know.
[11] QUBCO Attorney: The battery in SmartyPedo does not store charge the same way as the battery in SmartyFoot. Correct? Or, not correct?
[12] Jamie: I don't know.
[13] QUBCO Attorney: The battery in SmartyPedo does not provide the result of charging a cell phone. Correct? Or, not correct?
[14] Jamie: A battery is a battery.
[15] QUBCO Attorney: Correct? Or, not correct?
[16] Jamie: I don't know.
[17] QUBCO Attorney: You have a scholarship at the law school. Correct? Or, not correct?

[18] Jamie: Correct.
[19] QUBCO Attorney: Your scholarship is provided by JUGCO. Correct? Or, not correct?
[20] Jamie: That makes no difference.
[21] QUBCO Attorney: Thank you. No further questions.

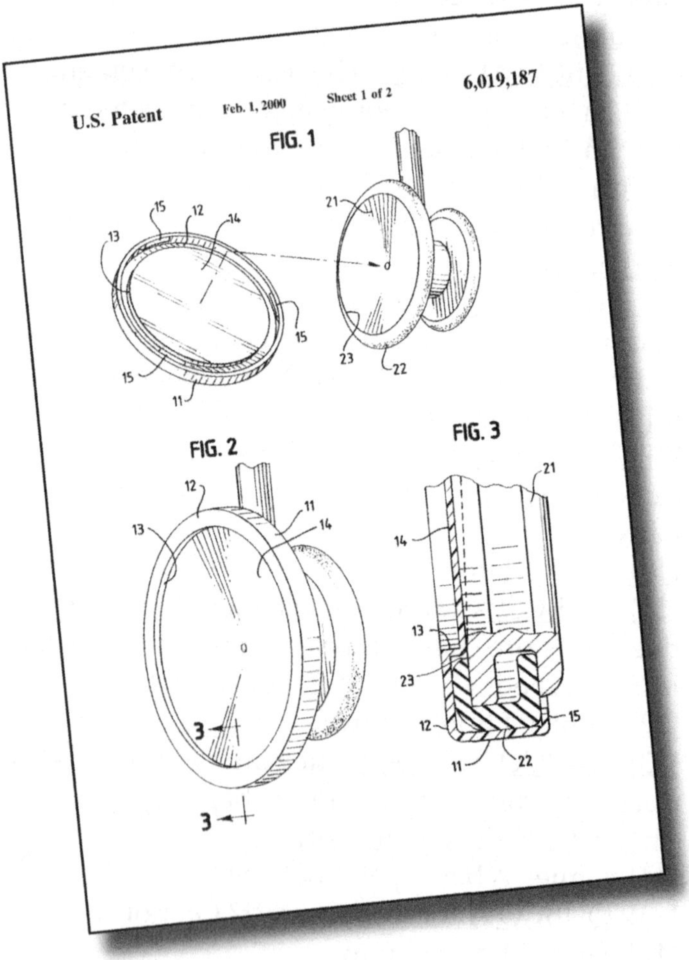

Deposition from the defendant's witness
Questioned by defendant's council
(Story for direct testimony)

[1] QUBCO Attorney: Please tell us your name and spell it for the record.
[2] Harley: Harley H A R L E Y.
[3] QUBCO Attorney: What is your occupation?
[4] Harley: I am a battery designer with thirty years experience, now an independent contractor.
[5] QUBCO Attorney: What types of batteries have you designed?
[6] Harley: All kinds from giant storage batteries big as a room to tiny batteries used in hearing aids.
[7] QUBCO Attorney: Are there important differences in how those batteries function?
[8] Harley: Many.
[9] QUBCO Attorney: For example, does the tiny battery in SmartyPedo function the same as the high capacity battery in SmartyFoot?
[10] Harley: The tiny battery in SmartyPedo uses different elements and different chemistry from the elements and chemistry in the high capacity battery in SmartyFoot. The tiny battery in SmartyPedo is in the class called primary batteries and the high capacity batter in SmartyFoot is in the class called secondary batteries.
[11] QUBCO Attorney: Let me be clear. Did you say that the SmartyPedo battery functions differently from the SmartyFoot battery?
[12] Harley: Yes, the SmartyPedo battery functions differently than the SmartyFoot battery.
[13] QUBCO Attorney: Thank you. No further questions.

Deposition from the defendant's witness
Questioned by plaintiff's council
(Story for <u>cross examination</u>)

<u>[1] JUGCO Attorney:</u> You said that the SmartyPedo battery functions differently from the SmartyFoot battery but they both have the same function of storing electrical energy. Correct? Or, not correct?
<u>[2] Harley:</u> They store the electrical energy in different ways.
<u>[3] JUGCO Attorney:</u> Correct, or not correct please. They both have the function of storing electrical energy. Correct? Or, not correct?
<u>[4] Harley:</u> Correct.
<u>[5] JUGCO Attorney:</u> They both store electrical energy as chemical energy so they both do this the same way. Correct? Or, not correct?
<u>[6] Harley:</u> Correct.
<u>[7] JUGCO Attorney:</u> They both produce the same result of providing electrical energy for electrical circuits. Correct? Or, not correct?
<u>[8] Harley:</u> Correct.
<u>[9] JUGCO Attorney:</u> You were once employed by HUGE GREEDY CO. Correct? Or, not correct?
<u>[10] Harley:</u> Correct.
<u>[11] JUGCO Attorney:</u> You lost that job because you falsified some results. Correct? Or, not correct?
<u>[12] Harley:</u> That has nothing to do with this.
<u>[13] JUGCO Attorney:</u> You admitted that you did falsify the results. Correct? Or, not correct?
<u>[14] Harley:</u> Correct.
<u>[15] JUGCO Attorney:</u> So you are an admitted liar. Correct? Or, not correct?

[16] Harley: That's not fair.
[17] JUGCO Attorney: Thank you. No further questions.

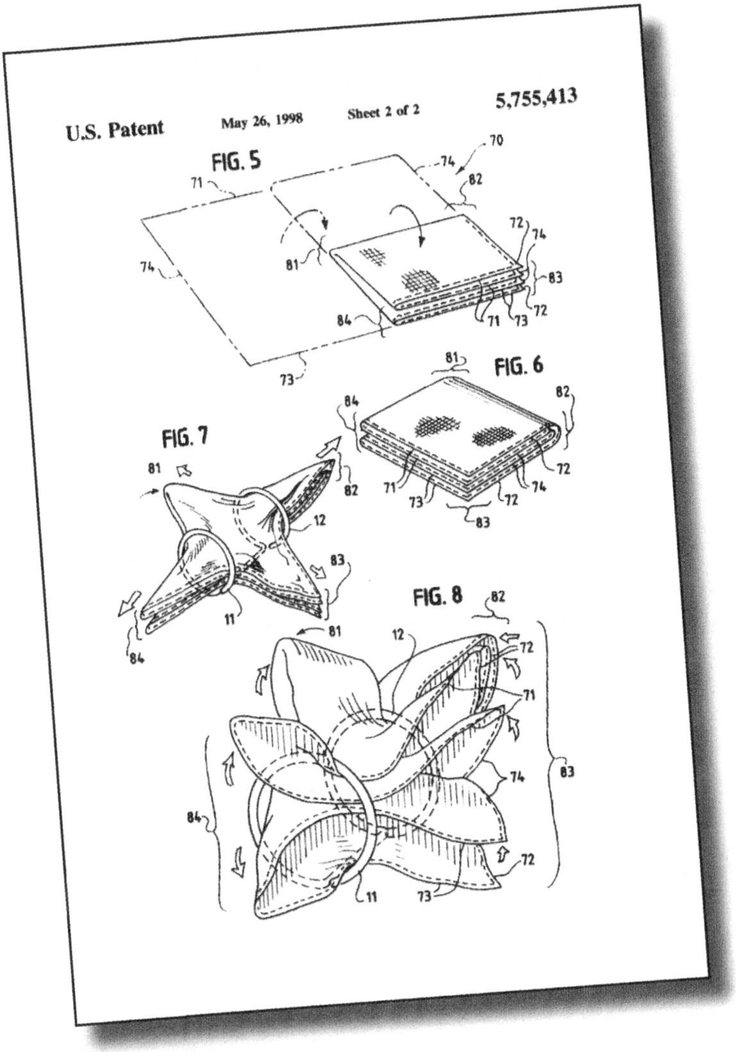

Can you identify some innovation wrongs which led to this trial?

Can you suggest some ways those innovation wrongs could have been put right?

If you were head of QUBCO what might you do to improve QUBCO's innovation rights?

If you were head of JUGCO, what might you do to improve JUGCO's innovation rights?

SmartyPedo Enable Trial

QUBCO invented SmartFoot but did not patent. JUGCO improved SmartFoot and patented SmartyFoot. QUBCO copied SmartyFoot so JUGCO sued QUBCO. QUBCO argued that SmartyFoot was not novel. JUGCO prevailed. The JUGCO SmartyFoot patent was valid.

QUBCO ceased making and selling SmartyFoot. Then QUBCO added a pedometer to SmartyFoot and began making and selling SmartyPedo, pat pend.

JUGCO sued QUBCO for infringing JUGCO's SmartyFoot patent. The jury found that QUBCO did infringe so QUBCO could not make nor sell SmartyPedo even though QUBCO had a valid patent on SmartyPedo.

Now SHELBYCO – remember Shelby – starts making and selling SmartPedo with a patent application pending. QUBCO makes and sells SmartPedo. When the SHELBYCO SmartPedo patent issues, SHELBYCO sues QUBCO for infringing the SmartPedo patent.

The SmartPedo patent application is almost word-for-word identical to QUBCO's SmartyPedo patent application, except that there is no mention of a battery in the claims nor in the written description.

QUBCO argues in defense that the SmartPedo patent is not valid because the patent application does not disclose how SmartPedo could work with no battery (not enabled).

In this case **the jury must decide** if a person having ordinary skill in the art could learn from the SHELBYCO patent how to make and use the invention.

Plaintiff's (SHELBYCO's) council and witness goals:

1) In the opening statement tell the jury that direct testimony by Riley will show that:

a) The SmartPedo patent is valid and QUBCO is infringing.

b) The invention is enabled because Riley had no problem making and using the invention.

2) In the direct testimony Riley, the plaintiff witness, convinces the jury of the points in 1) above.

3) In the cross examination discredit Sidney and get Sidney to admit that:

1) The direct testimony by Sidney is not trustworthy because Sidney did not test the circuit which Sidney claims would not work.

b) Sidney is not trustworthy because Sidney is paid to testify.

4) In the closing statement retell all the points in 1), 2), and 3) above and ask the jury to find for the plaintiff.

Defendant's (JUGCO's) council and witness goals:

1) In the opening statement tell the jury that direct testimony by Sidney will show that:

a) The SmartPedo patent is not valid because the circuit described in the written description would burn up quickly and thus the patent specification does not meet the requirement to enable the invention.

2) In direct testimony Sidney convinces jury of the points in 1) above.

3) In cross examination discredit Riley and get Riley to admit that the direct testimony by Riley is not trustworthy because Riley does not know what enable means in the context of patent law and does not know what is in the patent specification.

4) In the closing statement retell all the points in 1), 2), and 3) above and ask the jury to find for the defendant.

[Remember that the attorneys and witnesses can not change the "facts" on the next page and in the depositions. The attorneys can not invent new "facts" either. Please review the sections of the general introduction which explain the scenarios and trials.]

[Also remember that attorneys can object if a witness is asked to testify about something not directly known by the witness and can object if a witness testifies about something not directly known by the witness.]

These "facts" are not disputed by plaintiff and defense and are entered as evidence:

Jan Y1: QUBCO makes and sells SmartFoot no patent protection. *[Y1 = year one Y2 = year two Y3 = year three]*

Feb Y1: JUGCO files patent application and makes and sells SmartyFoot

Mar Y1: QUBCO makes and sells SmartyFoot

Jan Y2: SmartyFoot patent granted.

Feb Y2: Novelty trial with QUBCO in defense arguing SmartyFoot is not novel. JUGCO prevails in novelty trial.

Mar Y2: QUBCO adds pedometer to SmartyFoot to get SmartyPedo and makes and sells SmartyPedo, pat pend.

Apr Y2: Infringe Trial with QUBCO arguing in defense that SmartyPedo does not infringe SmartyFoot. JUGCO prevails.

May Y2: QUBCO's SmartyPedo patent application published

Nov Y2: SHELBYCO makes and sells SmartPedo pat pend.

Jan Y3: SmartyPedo patent grant to QUBCO, but since QUBCO prevailed in the infringe trial QUBCO can not nmake and sell SmartyPedo.

Feb Y3: SmartPedo patent grant to SHELBYCO.

Apr Y3: QUBCO makes and sells SmartPedo.

May Y3: SHELBYCO sues QUBCO for infringing SmartPedo with QUBCO arguing in defense that SmartPedo patent application did not enable claims in the SmartPedo patent.

The four depositions which follow are also "facts" which are entered as evidence.

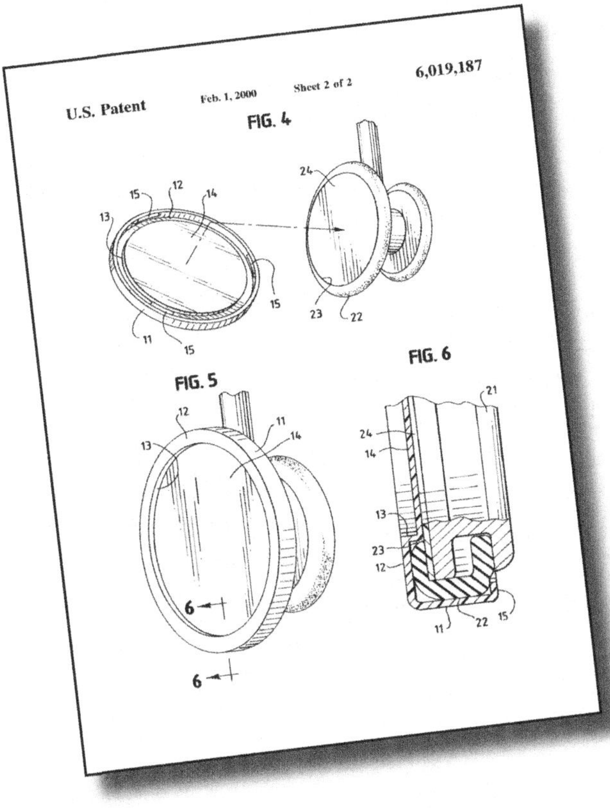

Deposition from the plaintiff's witness
Questioned by plaintiff's council
(Story for direct testimony)

[1] SHELBYCO Attorney: Please tell us your name and spell it for the record.
[2] Riley: Riley R I L E Y.
[3] SHELBYCO Attorney: What is your occupation?
[4] Riley: I am a junior at the University of Illinois studying electrical engineering.
[5] SHELBYCO Attorney: What is your connection with SmartPedo?
[6] Riley: Shelby, the president of SHELBYCO, is a friend of my parents and asked me to help with SmartPedo.
[7] SHELBYCO Attorney: What did you do for SmartPedo.
[8] Riley: I designed a circuit to make SmartPedo work.
[9] SHELBYCO Attorney: Was that hard?
[10] Riley: Nah, whipped it out in a few days.
[11] SHELBYCO Attorney: Did you need to do any experiments to design the circuit.
[12] Riley: No, just straight forward stuff.
[13] SHELBYCO Attorney: So, Shelby just gave you a list telling what SmartPedo must do?
[14] Riley: Yes, that's all I had to work with.
[15] SHELBYCO Attorney: Did Shelby tell you that this list was from the patent?
[16] Riley: I don't remember.
[17] SHELBYCO Attorney: Thank you. No further questions.

Deposition from the plaintiff's witness
Questioned by defendant's council
(Story for <u>cross examination</u>)

<u>[1] QUBCO Attorney:</u> Shelby asked you to design your circuit after the SmartPedo patent application was filed. Correct? Or, not correct?
<u>[2] Riley:</u> Correct.
<u>[3] QUBCO Attorney:</u> You designed your circuit without any reference to the SmartPedo patent application. Correct? Or, not correct?
<u>[4] Riley:</u> I did not need to.
<u>[5] QUBCO Attorney:</u> Correct? Or, not correct?
<u>[6] Riley:</u> Correct.
<u>[7] QUBCO Attorney:</u> You do not know if your circuit is enabled by the patent application. Correct? Or, not correct?
<u>[8] Riley:</u> What does "enabled" mean?
<u>[9] QUBCO Attorney:</u> The law says "The specification shall contain a written description of the invention, and of the manner and process of making and using it, in such full, clear, concise, and exact terms as to enable any person skilled in the art to which it pertains, or with which it is most nearly connected, to make and use the same" in section 112. You do not now know if the patent application does this. Correct? Or, not correct?
<u>[10] Riley:</u> Correct.
<u>[11] QUBCO Attorney:</u> SHELBYCO did not know how to make SmartPedo work when they asked you to design the circuit so SHELBYCO did not possess the invention when Shelby asked you to help. Correct? Or, not correct?
<u>[12] Riley:</u> I don't know.
<u>[13] QUBCO Attorney:</u> Thank you. No further questions.

Deposition from the defendant's witness
Questioned by defendant's council
(Story for direct testimony)

[1] QUBCO Attorney: Tell us your name and spell it for the record.
[2] Sidney: Sidney S I D N E Y.
[3] QUBCO Attorney: What is your occupation?
[4] Sidney: I am emeritus professor of electrical engineering at University of Illinois, Chicago.
[5] QUBCO Attorney: What is your specialty?
[6] Sidney: Circuits. Since I retired ten years ago I've been an expert witness at more than one patent trial each year.
[7] QUBCO Attorney: Very impressive. Do you see any connection between the circuit Riley designed and the SmartPedo patent?
[8] Sidney: Riley's circuit achieves the result described in the patent but Riley's circuit does this by a method very different from what is taught in the patent.
[9] QUBCO Attorney: How so?
[10] Sidney: The patent teaches a circuit which is appropriate only if there is a battery in the circuit. Without the battery the circuit taught in the patent will burn out in two miles of normal walking.
[11] QUBCO Attorney: Could a person of ordinary skill in the art get to Riley's circuit, or any working circuit, starting from what is taught in the patent?
[12] Sidney: No, one must start from scratch.
[13] QUBCO Attorney: In your expert opinion was SHELBYCO in possession of the invention at the time of filing the patent application?

[14] Sidney: No, they just copied the SmartyPedo patent and left out the battery without knowing that the circuit would burn out without the battery.
[15] QUBCO Attorney: Thank you. No further questions.

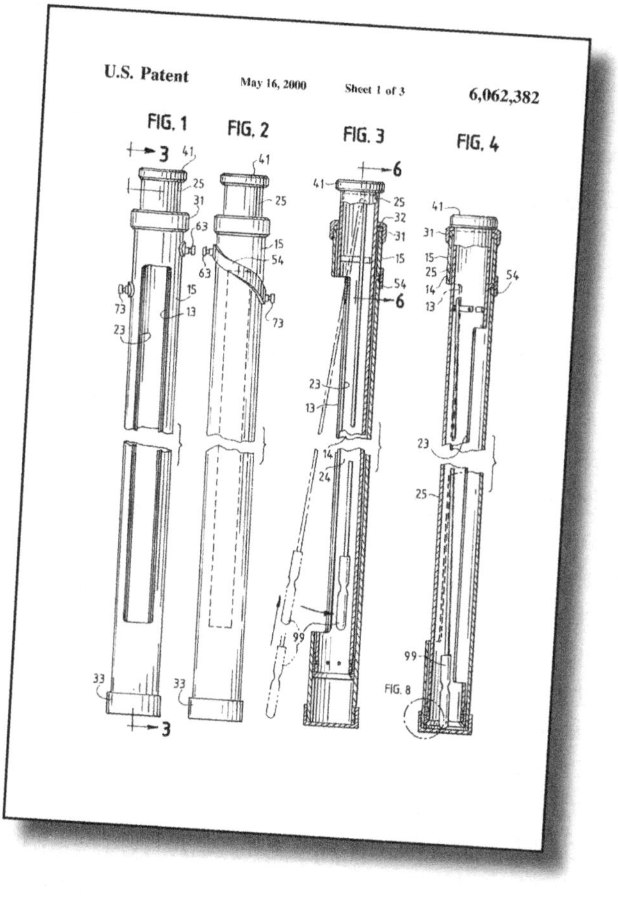

Deposition from the defendant's witness
Questioned by plaintiff's council
(Story for cross examination)

[1] SHELBYCO Attorney: Since you retired you have been hiring out as an expert witness for whichever side will pay you? Correct? Or, not correct?
[2] Sidney: I only testify the truth.
[3] SHELBYCO Attorney: Correct? Or, not correct?
[4] Sidney: Correct.
[5] SHELBYCO Attorney: QUBCO is paying you. Correct? Or, not correct?
[6] Sidney: Correct.
[7] SHELBYCO Attorney: You did not test the circuit in the SmartPedo patent. Correct? Or, not correct?
[8] Sidney: Correct.
[9] SHELBYCO Attorney: It is only your paid for opinion that the SmartPedo patent was copied from the SmartyPedo patent. Correct? Or, not correct?
[10] Sidney: Correct
[11] SHELBYCO Attorney: Many circuits could be designed to make SmartPedo work. Correct? Or, not correct?
[12] Sidney: Correct.
[13] SHELBYCO Attorney: It is only your paid for opinion that someone with ordinary skill in the art could not find one of those many circuits starting with the SmartPedo patent. Correct? Or, not correct?
[14] Sidney: Correct.
[15] SHELBYCO Attorney: Thank you. No further questions.

[If SHELBYCO prevails in this trial, then poor QUBCO:

> *1) can't make and sell SmartFoot,*
>
> *2) can't make and sell SmartyPedo, and*
>
> *3) can't make and sell SmartPedo*

even though QUBCO invented all the basic stuff.

Invention is not enough.

You must also:

> *1) understand innovation rights,*
>
> *2) understand how innovation rights can be lost,*
>
> *3) understand ways to make the most from innovation rights,*
>
> *4) understand ways to cooperate with competitors to make the most of innovation rights*

Can you identify some innovation wrongs which led to this trial?

Can you suggest some ways those innovation wrongs could have been put right?

If you were head of QUBCO what might you do to improve QUBCO's innovation rights?

If you were head of JUGCO, what might you do to improve JUGCO's innovation rights?

SmartPedo Employer Rights Trial

QUBCO invented SmartFoot but did not patent. JUGCO improved SmartFoot and patented SmartyFoot. QUBCO copied SmartyFoot so JUGCO sued QUBCO. QUBCO argued that SmartyFoot was not novel. JUGCO prevailed. The JUGCO SmartyFoot was valid.

QUBCO ceased making and selling SmartyFoot. Then QUBCO added a pedometer to SmartyFoot and began making and selling SmartyPedo.

JUGCO sued QUBCO for infringing JUGCO's SmartyFoot patent. The jury found that QUBCO did infringe so QUBCO could not make nor sell SmartyPedo even though QUBCO got a patent on SmartyPedo.

Then SHELBYCO – remember Shelby – started making and selling SmartPedo with a patent application pending. QUBCO copies SmartPedo. When the SHELBYCO SmartPedo patent issues, SHELBYCO sues QUBCO for infringing the SmartPedo patent. SHELBYCO prevails.

QUBCO now has nothing. JUGCO has SmartyFoot and also begins making and selling SmartPedo.

SHELBYCO sues JUGCO for infringing. JUGCO argues that JUGCO actually owns SmartPedo because Shelby's employment by JUGCO requires Shelby to assign SmartPedo to JUGCO.

In this case the **jury must decide**:

1) if SmartPedo is related to JUGCO's business.

2) if Shelby learned anything relevant to SmartPedo at JUGCO

Plaintiff's (SHELBYCO's) council and witness goals:

1) In the opening statement tell the jury that direct testimony by Shelby will show that:

a) The SmartPedo patent is valid and is infringed by JUGCO.
b) The SmartPedo patent is not related to JUGCO's business.
c) Shelby learned nothing relevant to the SmartPedo patent at JUGCO.

2) In the direct testimony Shelby, the plaintiff witness, convinces the jury of the points in 1) above.

3) In the cross examination discredit Rowan and get Rowan to admit that:

a) The direct testimony by Rowan is not trustworthy because Rowan gave mixed signals about SmartPedo being part of JUGCO's business.
b) Rowan is not trustworthy because Rowan has a grudge against Shelby.

4) In the closing statement retell all the points in 1), 2), and 3) above and ask the jury to find for the plaintiff.

Defendant's (JUGCO's) council and witness goals:

1) In the opening statement tell the jury that direct testimony by Rowan will show that:

a) The SmartPedo patent is the property of JUGCO because Shelby was obliged by employment contract to assign all rights to JUGCO.

b) JUGCO's employment contract did cover SmartPedo because SmartPedo was related to JUGCO's business and because Shelby learned about the SmartyFoot and SmartyPedo patents at JUGCO.

2) In direct testimony Rowan convinces jury of the points in 1) above.

3) In cross examination discredit Shelby and get Shelby to admit that:

a) The direct testimony by Shelby is not trustworthy because Shelby misunderstands employment contract and misunderstands JUGCO business.

b) Shelby is not trustworthy because Shelby has a grudge against Rowan;

4) In the closing statement retell all the points in 1), 2), and 3) above and ask the jury to find for the defendant.

[Also remember that attorneys can object if a witness is asked to testify about something not directly known by the witness and can object if a witness testifies about something not directly known by the witness.]

These "facts" are not disputed by plaintiff and defense and are entered as evidence:

Jan Y1: QUBCO makes and sells SmartFoot no patent protection*[Y1 = year one Y2 = year two Y3 = year three]*

Feb Y1: JUGCO files patent application and makes and sells SmartyFoot

Mar Y1: QUBCO makes and sells SmartyFoot

Jan Y2: SmartyFoot patent granted. JUGCO sues QUBCO for infringement.

Feb Y2: Novelty trial with QUBCO in defense arguing SmartyFoot is not novel. JUGCO prevails in novelty trial.

Mar Y2: QUBCO adds pedometer to SmartyFoot to get SmartyPedo and makes and sells SmartyPedo.

Apr Y2: Infringe trial with QUBCO arguing in defense that SmartyPedo does not infringe. QUBCO prevails and there is no valid patent on SmartyPedo. QUBCO offers to cross license with JUGCO so that both could make and sell SmartyPedo, but JUGCO refuses the offer.

May Y2: Shelby ends employment at JUGCO.

Jun Y2: SHELBYCO makes and sells SmartPedo Pat pend.

Feb Y3: SmartPedo Patent grant to SHELBYCO.

<u>Jun Y3</u>: JUGCO makes and sells SmartPedo.

<u>Jul Y3</u>: Shelby sues JUGCO for infringing the SmartPedo patent. JUGCO argus in defense that JUGCO owns SmartPedo patent because of <u>employer rights</u>.

The four depositions which follow are also "facts" which are entered as evidence.

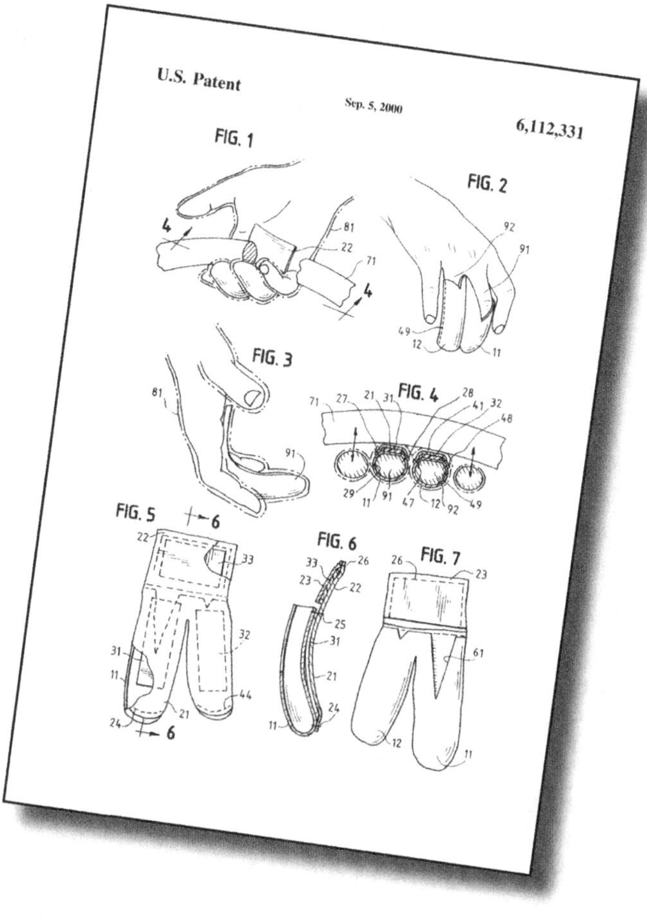

Deposition from the plaintiff's witness
Questioned by plaintiff's council
(Story for direct testimony)

[1] SHELBYCO Attorney: Please tell us your name and spell it for the record.
[2] Shelby: Shelby S H E L B Y.
[3] SHELBYCO Attorney: What is your connection with SmartPedo?
[4] Shelby: I learned about SmartyPedo from the SmartyPedo when the SmartyPedo patent application was published. I saw that removing the battery would make a less costly device which would not take as much energy away from the running or walking.
[5] SHELBYCO Attorney: What did you do?
[6] Shelby: Decided to start a business around SmartPedo. Filed a patent application and got going.
[7] SHELBYCO Attorney: Did you learn anything at JUGCO which contributed to the SmartPedo patent application?
[8] Shelby: Nothing. There is no work on smart products at JUGCO, just making clothing products. I learned everything from the SmartyPedo patent application.
[9] SHELBYCO Attorney: Didn't the JUGCO employment contract require you to assign SmartPedo rights to JUGCO?
[10] Shelby: No. SmartPedo has nothing to do with JUGCO business.
[11] SHELBYCO Attorney: How do you know that?
[12] Shelby: QUBCO offered to license SmartyPedo to JUGCO and JUGCO refused the

offer, showing that JUGCO had no interest in this type of product.

[13] SHELBYCO Attorney: Thank you. No further questions.

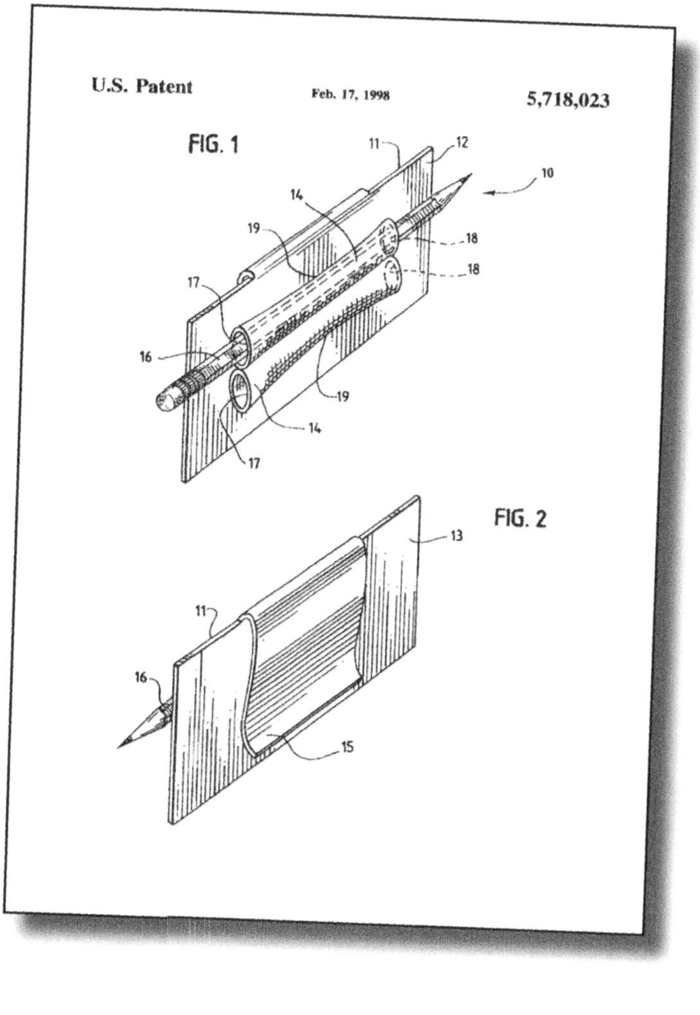

Deposition from the plaintiff's witness
Questioned by defendant's council
(Story for cross examination)

[1] JUGCO Attorney: You had a short romantic fling with Rowan. Correct? Or not Correct?
[2] Shelby: That's irrelevant.
[3] JUGCO Attorney: Correct? Or not Correct?
[4] Shelby: Correct.
[5] JUGCO Attorney: When that fling ended badly you said loudly in front of co-workers that Rowan is a jerk and you'll get revenge. Correct? Or not Correct?
[6] Shelby: Correct.
[7] JUGCO Attorney: You remember an email sent to all employees by Rowan that JUGCO intended to do more sport related products? Correct? Or not Correct?
[8] Shelby: Correct.
[9] JUGCO Attorney: SmartPedo is a sport related product. Correct? Or not Correct?
[10] Shelby: Correct.
[11] JUGCO Attorney: Thank you. No further questions.

Deposition from the defendant's witness
Questioned by defendant's council
(Story for direct testimony)

[1] JUGCO Attorney: Tell us your name and spell it for the record please.
[2] Rowan: Rowan R O W A N.
[3] JUGCO Attorney: What is your connection with JUGCO.
[4] Rowan: I am the third generation head.
[5] JUGCO Attorney: What the business of JUGCO?
[6] Rowan: JUGCO is a third generation clothing manufacturer. Under my successor Emerson we moved toward highly innovative and high value added clothing products. I moved JUGCO to sports related products.
[7] JUGCO Attorney: What is JUGCO's interest in smart products.
[8] Rowan: We think that it is time to combine smart products and clothing products. SmartyFoot for example.
[9] JUGCO Attorney: Why did you turn down QUBCO's offer to licanse SmartyPedo?
[10] Rowan: The terms of the license agreement they offered were not acceptable. We hope that QUBCO will agree to better terms.
[11] JUGCO Attorney: Please tell us about the email you sent to all JUGCO employees about sport related products.
[12] Rowan: In that email I told all the employees that JUGCO was fully committed to innovative sport related products and asked that everyone be on the lookout for new product ideas.
[13] JUGCO Attorney: Thank you. No further questions.

Deposition from the defendant's witness
Questioned by plaintiff's council
(Story for cross examination)

[1] SHELBYCO Attorney: You had a brief flirtation with Shelby. Correct. Or not correct?
[2] Rowan: Nothing to it.
[3] SHELBYCO Attorney: Correct? or not Correct?
[4] Rowan: Correct.
[5] SHELBYCO Attorney: When Shelby denounced you in front of coworkers and said that she'd get revenge, you were heard to say that Shelby would regret that. Correct? or not Correct?
[6] Rowan: I only meant that Shelby would be sorry for the dissonance created by the harsh words.
[7] SHELBYCO Attorney: Correct? or not Correct?
[8] Rowan: Correct.
[9] SHELBYCO Attorney: Your email about sports related products did not mention smart products. Correct? or not Correct?
[10] Rowan: Correct.
[11] SHELBYCO Attorney: There is now no work underway at JUGCO related to smart products. Correct? or not Correct?
[12] Rowan: Correct.
[13] SHELBYCO Attorney: Thank you. No further questions.

Can you identify some innovation wrongs which led to this trial?

Can you suggest some ways those innovation wrongs could have been put right?

If you were head of QUBCO what might you do to improve QUBCO's innovation rights?

If you were head of JUGCO, what might you do to improve JUGCO's innovation rights?

SmartyPedo License Arbitration

Our first starting point had QUBCO making and selling SmartFoot with no patent pending, and that led to all kinds of trouble with possibly JUGCO owning everything and QUBCO owning nothing.

This time let's start with QUBCO making and selling SmartFoot <u>with</u> patent pending status. JUGCO still makes improvements also with patent pending status.

When the QUBCO and the JUGCO patent applications are published, as patent applications are, QUBCO and JUGCO see that they have a dicey situation. JUGCO can't make and sell SmartyFoot without infringing a SmartFoot patent and QUBCO can't incorporate the JUGCO improvements without infringing a SmartyFoot patent.

They agree to cross license: JUGCO gets rights to SmartFoot so that JUGCO can make and sell SmartyFoot and QUBCO gets rights to SmartyFoot so QUBCO can make and sell SmartyFoot.

The license specifies that any improvements in either of the inventions will be included in the licensed rights. The license also specifies that any disputes must be settled by mandatory arbitration and further specifies that the arbitrator hold a mini-trial to hear the arguments and evidence.

The problem comes when QUBCO adds the pedometer and makes and sells SmartyPedo and argues that this is not covered by the license. JUGCO invokes the arbitration clause in the license.

The **arbitrator must decide** if SmartyPedo is – or is not – an improvement of either SmartFoot or SmartyFoot.

Plaintiff's (JUGCO's) council and witness goals:

1) In the opening statement tell the arbitrator that direct testimony by Rowan will show that the SmartyPedo patent has all the elements in the SmartyFoot patent and is an improvement in the ordinary meaning of invention.

2) In the direct testimony Rowan convinces the arbitrator of the points in 1) above.

3) In the cross examination discredit Brook and get Brook to admit that Brook's definition of "invention" is too narrow used only because Brook stands to gain if the arbitrator rules in favor of QUBCO.

4) In the closing statement retell all the points in 1), 2), and 3) above and ask the arbitrator to find for the plaintiff.

Defendant's (QUBCO's) council and witness goals:

1) In the opening statement tell the arbitrator that direct testimony by Brook will show that improving an invention means improving something in the existing claims in a patent, and SmartPedo does not improve anything in existing claims in the SmartyPedo patent.

2) In direct testimony Brook convinces arbitrator of the points in 1) above.

3) In cross examination discredit Rowan and get Rowan to admit that Rowan uses a too broad definition of invention only because Rowan stands to gain if the arbitrator decides in favor of JUGCO;

4) In the closing statement retell all the points in 1), 2), and 3) above and ask the arbitrator to find for the defendant.

[Remember that the attorneys and witnesses can not change the "facts" on the next page and in the depositions. The attorneys can not invent new "facts" either. Please review the sections of the general introduction which explain the scenarios and trials.]

[Also remember that attorneys can object if a witness is asked to testify about something not directly known by the witness and can object if a witness testifies about something not directly known by the witness.]

These "facts" are not disputed by plaintiff and defense and are entered as evidence:

Jan Y1: QUBCO files patent application and makes and sells SmartFoot no patent protection. *[Y1 = year one, Y2 = year 2]*

Feb Y1: JUGCO files patent application and makes and sells SmartyFoot

Jul Y1: QUBCO SmartFoot patent application published. JUGCO SmartyFoot patent application published.

Aug Y1: QUBCO and JUGCO cross license SmartFoot and SmartyFoot. The license agreement specifies that any improvement in either invention will be included in the license agreement.

Mar Y2: QUBCO makes and sells SmartyPedo, pat pend.

Apr Y2: JUGCO invokes license **Arbitration**

Deposition from the plaintiff's witness
Questioned by plaintiff's council
(Story for direct testimony)

[1] JUGCO Attorney: Please tell us your name and spell it for the record.
[2] Rowan: Rowan R O W A N.
[3] JUGCO Attorney: What is your connection with JUGCO?
[4] Rowan: I am the third generation head succeeding Emerson who passed. Before that I was Emerson's assistant for many years.
[5] JUGCO Attorney: Do you have experience in all parts of the business?
[6] Rowan: Yes, everything from cleaning the bathrooms to doing long-term planning.
[7] JUGCO Attorney: Why do you think that QUBCO is violating the license agreement?
[8] Rowan: The license agreement says that any improvements in the invention are included in the license. SmartyPedo is an improvement in the invention because it adds another function to the existing product.
[9] JUGCO Attorney: Are you equating the term "invention" in the license agreement to the product?
[10] Rowan: Of course. Nothing else makes sense. QUBCO can't hide behind some narrow definition of invention. The license is about business so the word must mean something which persons in business understand.
[11] JUGCO Attorney: Are you saying that persons in business do understand "invention" to mean the product.

[12] Rowan: Yes! I've spent many years making written and oral agreements with our vendors and customers. Knowing how they use language is essential for my success.
[13] QUBCO Attorney: In your experience making agreements with others in business, how are meanings of words in agreements treated?
[14] Rowan: In my experience words have their ordinary meaning unless otherwise specified. If QUBCO wanted the word "invention" to have a narrow meaning, then QUBCO should have had that narrow meaning specified in the license agreement.
[15] JUGCO Attorney: Thank you. No further questions.

Deposition from the plaintiff's witness
Questioned by defendant's council
(Story for cross examination)

[1] QUBCO Attorney: The license agreement is about patent rights. Correct? Or not correct?
[2] Rowan: Correct.
[3] QUBCO Attorney: So the terms in the license agreement should have their meaning in patent law.
[4] Rowan: I don't think so.
[5] QUBCO Attorney: The income of JUGCO will increase if JUGCO can make and sell SmaryPedo. Correct? Or not correct?
[6] Rowan: Correct.
[7] QUBCO Attorney: So you want to give the word "invention" a meaning different from it's meaning in patent law so JUGCO's income can increase. Correct? Or not correct?
[8] Rowan: I want the word to mean want it normally means.
[9] QUBCO Attorney: Correct? Or not correct?
[10] Rowan: Not corect.
[11] QUBCO Attorney: Thank you. No further questions.

[12] Deposition from the defendant's witness
Questioned by defendant's council
(Story for direct testimony)

[1] QUBCO Attorney: Please tell us your name and spell it for the record.
[2] Brook: Brook B R O O K.
[3] QUBCO Attorney: What is your connection with QUBCO?
[4] Brook: I am co-owner and I invent our products.
[5] QUBCO Attorney: Do you have much experience with patents?
[6] Brook: Lots. I am responsible for many patents. I've worked with many patent professionals. I ask lots of questions and get good advice about patent law. Understanding patent law is crucial to my success.
[7] QUBCO Attorney: What does the word "invention" mean?
[8] Brook: An "invention" is that which is claimed in a valid patent.
[9] QUBCO Attorney: Based on your experience, why do you say that SmartyPedo is not an improvement on either the SmartFoot invention or the SmartyFoot invention?
[10] Brook: To improve an invention one must improve something in one of the claims in a valid patent. SmartyPedo does not change any of the claims in either the SmartFoot patent or the SmartyFoot patent.
[11] QUBCO Attorney: Based on your experience, why do you say that SmartyPedo should not be included in what is covered by the license.

[12] Brook: The license is about patent rights, so the words mean what they mean in patent law. SmartyPedo does not change anything in the SmartFoot claims nor anything in the SmartyFoot claims. SmartyPedo is not included in the license.
[13] QUBCO Attorney: Are you saying that SmartyPedo is a new invention?
[14] Brook: Yes, new and not connected to the other inventions.
[15] QUBCO Attorney: Thank you. No further questions.

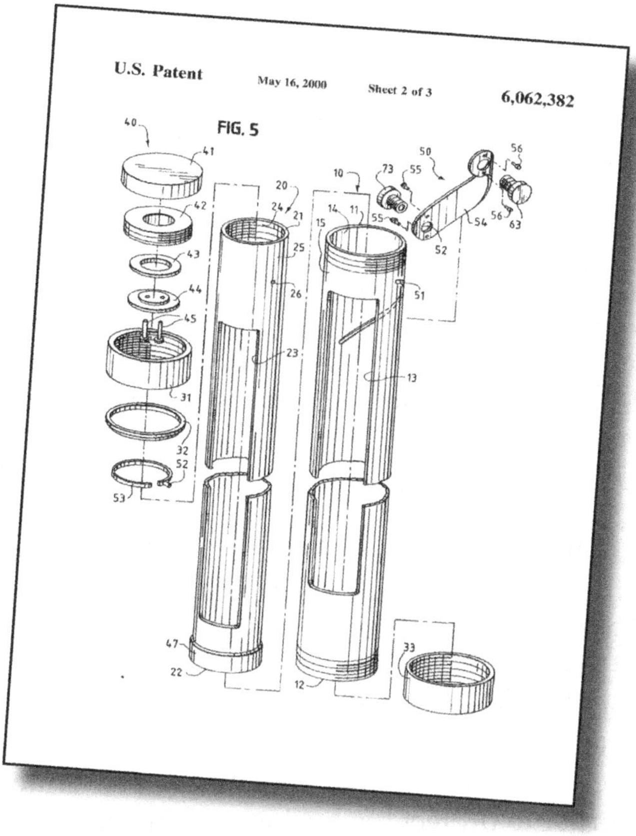

Deposition from the defendant's witness
Questioned by plaintiff's council
(Story for cross examination)

[1] JUGCO Attorney: The license agreement is about what JUGCO and QUBCO can make and sell. Correct? Or not correct?
[2] Brook: Correct.
[3] JUGCO Attorney: So, the license is about business. Correct? Or not correct?
[4] Brook: Not correct.
[5] JUGCO Attorney: You just said that an agreement about what QUBCO can make and sell is not about business. Correct? Or not correct?
[6] Brook: Correct.
[7] JUGCO Attorney: You are saying that making and selling a product is not business. Correct? Or not correct?
[8] Brook: Correct.
[9] JUGCO Attorney: If the license covers SmartyPedo, then JUGCO can make and sell SmartyPedo. Correct? Or not correct?
[10] Brook: The license does not cover SmartyPedo.
[11] JUGCO Attorney: Correct? Or not correct?
[12] Brook: Correct.
[13] JUGCO Attorney: When JUGCO makes and sells SmartyPedo QUBCO's income will go down. Correct? Or not correct?
[14] Brook: I don't know.
[15] JUGCO Attorney: You want the arbitrator to agree that the license is not about business and to agree with your narrow definition of "invention" in order to protect the business of QUBCO. Correct? Or not correct?

[16] Brook: You are putting too many words in my mouth.
[17] JUGCO Attorney: Thank you. No further questions.

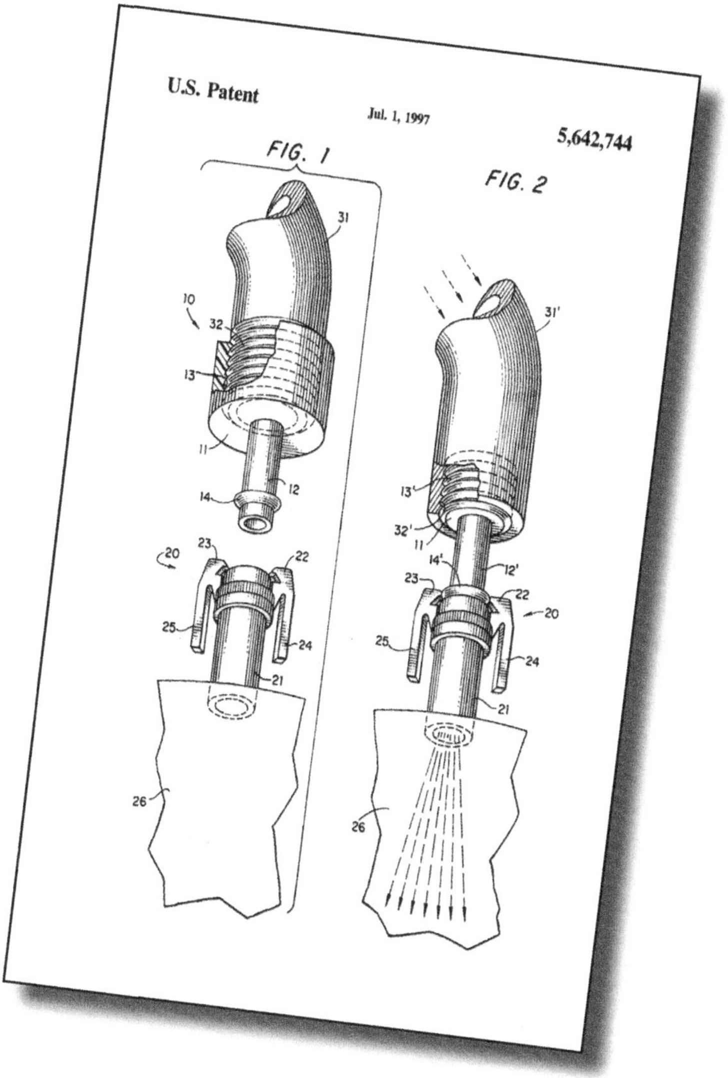

Can you identify some innovation wrongs which led to this trial?

Can you suggest some ways those innovation wrongs could have been put right?

If you were head of QUBCO what might you do to improve QUBCO's innovation rights?

If you were head of JUGCO, what might you do to improve JUGCO's innovation rights?

SmartyPedo Joint Venture Arbitration

Many bad things happened after the first novelty trial.

QUBCO, which made the key inventions, could end up with nothing. JUGCO which made key improvements could end up with nothing. SHELBYCO could have poached both SmartyFoot and SmartyPedo.

Could those bad things have been avoided?

Maybe. But, then new bad things can happen. Here's another fresh start:

QUBCO created, made, and sold SmartFoot which is a shoe insert using piezoelectric elements to harvest energy of walking and running to charge a battery. That battery is then used at night to charge smart phones and like devices.

QUBCO did not file a patent application for SmartFoot. JUGCO copied SmartFoot and made improvements to make and sell SmartyFoot which worked better, lasted longer, fit more shoes, and cost much less.

QUBCO copied and made and sold SmartyFoot. JUGCO did file a patent application covering the SmartyFoot improvements. When the SmartyFoot patent grant issued, JUGCO sued QUBCO for infringing the SmartyFoot patent.

This time QUBCO and JUGCO decided to settle with JUGCO giving QUBCO license to make and sell SmartyFoot and with QUBCO and JUGCO agreeing to a joint venture contract so that they could make innovations which would contribute to QUBCO's business and to JUGCO's business.

QUBCO would get access to JUGCO's experience in manufacturing and market knowledge and to get access to JUGCO's innovation process. JUGCO would get access to QUBCO's inventiveness and advanced technologies.

Sounds like a good deal, but innovation joint venture contracts are very slippery because the very nature of innovation makes it impossible to guess what the results of the joint venture will be leaving many opportunities for dispute.

Dispute there was. QUBCO invented SmartyPedo which added a pedometer to SmartyFoot and claimed that SmartyPedo is not covered by the joint venture agreement.

The joint venture contract which JUGCO and QUBCO agreed to, and signed, specifies that any disputes about the contract must be settled by mandatory arbitration.

The joint venture contract further specifies that the arbitrator hold a mini-trial to hear the arguments and evidence.

[Surprise! It just happens that the form of the mini-trial specified in the joint venture contract is the same form as our mock trial.]

[This is a good example of how innovation rights involve more than patent rights. Innovation rights can be gained, and can be lost, in the writing and adjudication of contracts.]

This dispute about what the joint venture covers will be adjudicated by an arbitrator, who will hold a "trial" to understand the facts of the case.

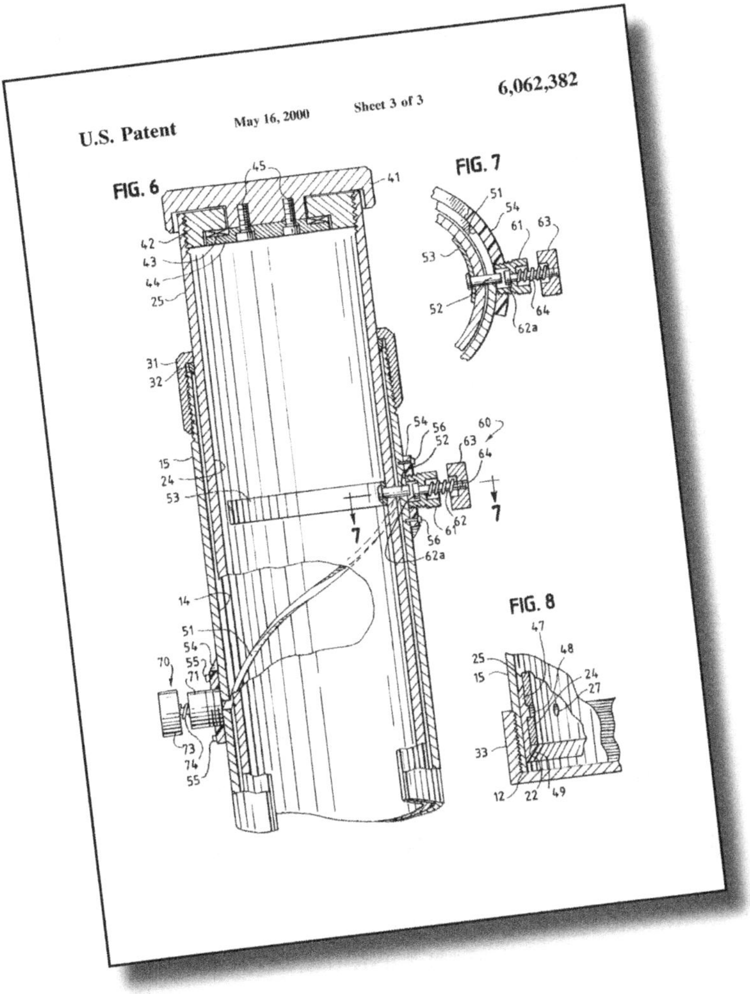

In this "trial" **the arbitrator must decide** if SmartyPedo is, or is not, an innovation covered by the joint venture contract.

[This is very tricky. Since the joint venture contract is about creating innovations, it can not be known at the time the contract is written what will be created. So, the contract can not specify precisely what are, and are not, outcomes covered by the contract.

This lack of specificity leaves plenty of room for argument. Thus, especially here, the outcome of the arbitration will depend on which side is the most persuasive.]

Plaintiff's (JUGCO's) council and witness goals:

1) In the <u>opening statement</u> tell the arbitrator that direct testimony by Rowan will show that since SmartyPedo is a shoe insert, SmartyPedo is related to JUGCO's business and thus is covered by joint venture agreement.

2) In the <u>direct testimony</u> Rowan convinces the arbitrator of the points in 1) above.

3) In the <u>cross examination</u> discredit Quinn and get Quinn to admit that SmartyPedo uses all the improvements JUGCO made to SmartFoot and therefore is related to JUGCO's business.

4) In the <u>closing statement</u> retell all the points in 1), 2), and 3) above and ask the arbitrator to find for the plaintiff.

Defendant's (QUBCO's) council and witness goals:

1) In the opening statement tell the arbitrator that direct testimony by Quinn will show that SmartyPedo is a smart product, which is QUBCO's business and smart products are not part of JUGCO's business.

2) In direct testimony Quinn convinces arbitrator of the points in 1) above.

3) In cross examination discredit Rowan and get Rowan to admit that JUGCO has no capacity for smart products;

4) In the closing statement retell all the points in 1), 2), and 3) above and ask the arbitrator to find for the defendant.

[Remember that the attorneys and witnesses can not change the "facts" on the next page and in the depositions. The attorneys can not invent new "facts" either. Please review the sections of the general introduction which explain the scenarios and trials.]

[Also remember that attorneys can object if a witness is asked to testify about something not directly known by the witness and can object if a witness testifies about something not directly known by the witness.]

These "facts" are not disputed by plaintiff and defense and are entered as evidence:

Jan Y1: QUBCO makes and sells SmartFoot no patent protection. *[Y1 = year one, Y2 = year 2]*

Feb Y1: JUGCO files patent application and makes and sells SmartyFoot

Mar Y1: QUBCO makes and sells SmartyFoot

Jan Y2: SmartyFoot patent granted. JUGCO sues QUBCO for infringement.

Feb Y2: SmartyFoot novelty trial with QUBCO in defense arguing SmartyFoot is not novel. JUGCO and QUBCO settle with JUGCO licensing to QUBCO rights to make and sell SmartyCoat. And, as part of the settlement JUGCO and QUBCO agree to a joint venture to develop new products. JUGCO gets access to Brook's extraordinary inventive abilities. QUBCO gets access to JUGCO's extraordinary group innovation process.

Mar Y2: QUBCO adds pedometer to SmartyFoot to get SmartyPedo and applies for patent without mentioning JUGCO.

Apr Y2: JUGCO accuses QUBCO of violation of their joint venture contract. The contract calls for mandatory arbitration in the form of a trial.

May Y2: **Arbitration**

Deposition from the plaintiff's witness
Questioned by plaintiff's council
(Story for direct testimony)

[1] JUGCO Attorney: Please state your name and spell it for the record.
[2] Rowan: Rowan R O W A N
[3] JUGCO Attorney: What is your occupation?
[4] Rowan: I am the third generation head of JUGCO, succeeding my parent Emerson.
[5] JUGCO Attorney: What is JUGCO's business?
[6] Rowan: For three generations we make and sell clothing.
[7] JUGCO Attorney: Has there been a change in the business while you have been involved?
[8] Rowan: Yes, a big change. Under Emerson our competitors – especially overseas – were taking our business with low cost manufacturing.
[9] JUGCO Attorney: What did you do?
[10] Rowan: Emerson started an internal innovation process. When we see a good new product, all our workers contribute ideas to make enough improvements in that new product so that we have our own innovative product which does not infringe any existing patents, is better than any existing products, and costs less than existing products.
[11] JUGCO Attorney: Is this what happened to lead to SmartyFoot?
[12] Rowan: Yes.
[13] JUGCO Attorney: Did SmartyFoot change the focus of your business?
[14] Rowan: Yes, certainly. This was our second smart clothing product.
[15] JUGCO Attorney: How did you get involved with QUBCO?

[16] Rowan: QUBCO copied SmartyFoot thinking that they could invalidate our SmartyFoot patent because their SmartFoot anticipated some parts of SmartyFoot. We sued QUBCO for infringement. If we lost that suit, then no one would have any patent protection for SmatyFoot and some huge greedy co would take all the business. So, we cross-licensed with QUBCO and entered into the joint venture contract.

[17] JUGCO Attorney: What do you get from the joint venture?

[18] Rowan: We get access to QUBCO's advanced technologies and QUBCO gets access to our extraordinary innovation process.

[19] JUGCO Attorney: Why do you think that the joint venture contract covers SmartyPedo?

[20] Rowan: It is a shoe insert using the manufacturing innovations claimed in the SmartyFoot patent. Shoe inserts are areas of overlap of the business of JUGCO and QUBCO. It's a slam dunk!

[21] JUGCO Attorney: Thank you. No further questions.

Deposition from the plaintiff's witness
Questioned by defendant's council
(Story for cross examination)

[1] QUBCO Attorney: Your first smart product was SmartyCoat and the "smarts" in SmartyCoat came from QUBCO. Correct? Or not correct?
[2] Rowan: Correct.
[3] QUBCO Attorney: Your second smart product is SmartyFoot and the smarts in SmartyFoot came from QUBCO. Correct? Or not correct?
[4] Rowan: Correct.
[5] QUBCO Attorney: JUGCO only contributed manufacturing know-how to those products. Correct? Or not correct?
[6] Rowan: Correct.
[7] QUBCO Attorney: All the manufacturing know-how in SmartyPedo is disclosed in the SmartyFoot patent. Correct? Or not correct?
[8] Rowan: Correct.
[9] QUBCO Attorney: So, your clothing business only involves smart products derived from QUBCO and you contributed nothing new to SmartyPedo. Correct? Or not correct?
[10] Rowan: Correct.
[11] QUBCO Attorney: So, there's nothing joint about SmartyPedo. Correct? Or not correct?
[12] Rowan: That's for the arbitrator to decide.
[13] QUBCO Attorney: Thank you. No further questions.

Deposition from the defendant's witness
Questioned by defendant's council
(Story for direct testimony)

[1] QUBCO Attorney: Please state your name and spell it for the record.
[2] Quinn: Quinn Q U I N N.
[3] QUBCO Attorney: What is your occupation?
[4] Quinn: I am co-owner and president of QUBCO.
[5] QUBCO Attorney: What is the business of QUBCO.
[6] Quinn: We invent, make, and sell smart products. These are products which provide useful information about a person's life.
[7] QUBCO Attorney: How did SmartFoot come about?
[8] Quinn: Our smart products need power, so we invented a way to harvest energy of walking to charge a battery. The battery then can charge up your cell phone and our smart products at night.
[9] QUBCO Attorney: How did you get involved with JUGCO?
[10] Quinn: We rushed SmartFoot to market because it is outside our smart product business. It didn't work very well and didn't fit many shoes. JUGCO added their manufacturing know-how and patented SmartyFoot. Since SmartFoot anticipated SmartyFoot we started making and selling the same product.
[11] QUBCO Attorney: What happened?
[12] Quinn: JUGCO sued us for infringing their SmartyFoot patent. We might have invalidated their patent but did not want to spend the money. By cross licensing so that JUGCO has rights to

SmartFoot and QUBCO has rights to SmartyFoot we have a cash cow to support our smart product business.

[13] QUBCO Attorney: Why didn't you patent SmartFoot?

[14] Quinn: We should have. Then we could have licensed it to JUGCO and let them do all the work making and selling with QUBCO getting royalties to support the smart product business.

[15] QUBCO Attorney: Where did SmartyPedo come from?

[16] Quinn: The idea is to have a pedometer in the shoe with a bluetooth connection to your smart phone to tell you all sorts of information about how you are walking or running. The easy way to get it on the market was to incorporate it into SmartyFoot to make SmartyPedo.

[17] QUBCO Attorney: Doesn't selling SmartyFoot mean that your business overlaps JUGCO's?

[18] Quinn: No of course not. That's like saying that because the Art Institute of Chicago sells scarves in their gift shop they are in the scarf business. That's silly.

[19] QUBCO Attorney: Isn't SmartyPedo a joint product with JUGCO because it uses manufacturing know-how disclosed in the SmartyFoot patent?

[20] Quinn: That's like saying that my computer is a joint product with Remington Typewriter because my computer has a qwerty keyboard. Baloney.

[21] QUBCO Attorney: Thank you. No further questions.

Deposition from the defendant's witness
Questioned by plaintiff's council
(Story for cross examination)

[1] JUGCO Attorney: You make and sell SmartyFoot. Correct? Or not correct?
[2] Quinn: Correct.
[3] JUGCO Attorney: The SmartyFoot patent covers manufacturing steps contributed by JUGCO. Correct? Or not correct?
[4] Quinn: Correct.
[5] JUGCO Attorney: The SmartyPedo patent application also incorporates the manufacturing steps in the SmartyFoot patent. Correct? Or not correct?
[6] Quinn: Correct
[7] JUGCO Attorney: So, you make and sell SmartyFoot which is a product identical to a product made and sold by JUGCO and your SmartyPedo patent application includes manufacturing steps first patented by JUGCO. Correct? Or not correct?
[8] Quinn: Correct.
[9] JUGCO Attorney: So, SmartyPedo is in an area of business which overlaps JUGCO's business and SmartyPedo joins the pedometer with SmartyFoot. This is exactly what the joint venture contract covers. Correct? Or not correct?
[10] Quinn: That's for the arbitrator to decide.
[11] JUGCO Attorney: Thank you. No further questions.

Can you identify some innovation wrongs which led to this trial?

Can you suggest some ways those innovation wrongs could have been put right?

If you were head of QUBCO what might you do to improve QUBCO's innovation rights?

If you were head of JUGCO, what might you do to improve JUGCO's innovation rights?

SmartyPedo Inequitable Conduct Trial

QUBCO invented SmartFoot but did not patent. JUGCO improved SmartFoot and patented SmartyFoot. QUBCO copied SmartyFoot so JUGCO sued QUBCO. Rather than go to trial, QUBCO and JUGCO settled the case with JUGCO licensing to QUBCO rights to make and sell SmartyCoat. JUGCO and QUBCO also agree to a joint venture to develop new products.

QUBCO added a pedometer to SmartyFoot and patented SmartyPedo. After an arbitrator ruled that SmartyPedo was not covered by the joint venture agreement, JUGCO went ahead and copied SmartyPedo anyway.

QUBCO sues JUGCO for infringement. In the trial JUGCO argues that the QUBCO SmartyPedo patent is not valid because QUBCO did not disclose to the USPTO the joint venture contract, which is inequitable conduct with the USPTO.

A court can decide that your patent is not valid if you engaged in inequitable conduct before the USPTO. For example it can be inequitable conduct to withhold information relevant to the patentability of your invention.

In our case **the jury must decide**:

1) if – or if not – QUBCO withheld information which might have led to rejection of claims; and

2) if QUBCO intended to deceive the USPTO to get an undeserved patent.

[Intention is a very slippery issue.]

Plaintiff's (QUBCO's) council and witness goals:

1) In the opening statement tell the jury that direct testimony by Quinn will show that:

a) The SmartyPedo patent is valid and JUGCO is infringing.
b) There was no inequitable conduct because SmartyPedo is not related to JUGCO's business and has no connection to the joint venture agreement.

2) In the direct testimony Quinn convinces the jury of the points in 1) above.

3) In the cross examination discredit Rowan and get Rowan to admit that:

a) The arbitrator ruled that SmartyPedo is not covered by the joint venture agreement.
b) Rowan claims to be able to read minds.

4) In the closing statement retell all the points in 1), 2), and 3) above and ask the jury to find for the plaintiff.

Defendant's (JUGCO's) council and witness goals:

1) In the opening statement tell the jury that direct testimony by Rowan will show that:

a) The SmartyPedo patent is not valid because QUBCO did not inform the USPTO about the joint agreement.
b) An arbitrator can not make decisions about patentability only the USPTO can.
c) QUBCO used the joint venture agreement as bait to get a license to improvements JUGCO made to SmartFoot and those improvements are in the claims of the SmartyPedo patent.
d) The USPTO had a right to know about the joint venture agreement.

2) In direct testimony Rowan convinces jury of the points in 1) above.

3) In cross examination discredit Quinn and get Quinn to admit that QUBCO did not deal honestly with JUGCO.

4) In the closing statement retell all the points in 1), 2), and 3) above and ask the jury to find for the defendant.

[Also remember that attorneys can object if a witness is asked to testify about something not directly known by the witness and can object if a witness testifies about something not directly known by the witness.]

These "facts" are not disputed by plaintiff and defense and are entered as evidence:

Jan Y1: QUBCO makes and sells SmartFoot no patent protection*[Y1 = year one Y2 = year two Y3 = year three]*

Feb Y1: JUGCO files patent application and makes and sells SmartyFoot

Mar Y1: QUBCO makes and sells SmartyFoot

Jan Y2: SmartyFoot patent granted. JUGCO sues QUBCO for infringement.

Feb Y2: Novelty trial with QUBCO in defense arguing SmartyFoot is not novel. JUGCO and QUBCO settle with JUGCO licensing to QUBCO rights to make and sell SmartyCoat. JUGCO and QUBCO also agree to a joint venture to develop new products. JUGCO gets access to Brook's extraordinary inventive abilities. QUBCO gets access to JUGCO's extraordinary group innovation process.

Mar Y2: QUBCO adds pedometer to SmartyFoot to get SmartyPedo and applies for patent.

Apr Y2: JUGCO accuses QUBCO of violation of joint venture contract. Contract calls for mandatory arbitration in the form of a trial.

May Y2: Contract arbitration trial. QUBCO prevails. The joint venture contract is voided.

Feb Y3: SmartyPedo patent granted to QUBCO.

Apr Y3: JUGCO makes and sells SmartyPedo

May Y3: QUBCO sues JUGCO for infringing SmartyPedo patent.

Jun Y3: **Inequitable Conduct Trial** with JUGCO arguing in defense that QUBCO engaged in inequitable conduct by not informing the USPTO about the joint venture contract.

The four depositions which follow are also "facts" which are entered as evidence.

Deposition from the plaintiff's witness
Questioned by plaintiff's council
(Story for direct testimony)

[1] QUBCO Attorney: Please tell us your name and spell it for the record.
[2] Quinn: Quinn Q U I N N.
[3] QUBCO Attorney: What is your occupation?
[4] Quinn: I am co-owner and President of QUBCO. We invent, make, and sell smart products.
[5] QUBCO Attorney: Was SmartFoot one of those smart products?
[6] Quinn: No, it was an auxiliary product. SmartFoot harvests walking energy to charge a battery which then can be used to charge our smart products at night.
[7] QUBCO Attorney: Why did you settle the infringement suit which JUGCO brought against you?
[8] Quinn: We expected to win the suit but then there would be no patent so some huge greedy company would make all the profit on our invention. By settling we could keep some control over the product and coutinue to have an income stream from the product.
[9] QUBCO Attorney: Why did you agree to the joint venture contract?
[10] Quinn: We thought that we would benefit from the great innovation process at JUGCO. But, we found that their innovations only concerned clothing manufacture and was no use for our smart products.
[11] QUBCO Attorney: Did the joint venture have any connection to SmartyPedo?

<u>[12] Quinn:</u> Not at all. SmartyPedo is a smart product not a clothing product. It just happens to fit well with SmartyFoot. So, since we have license rights to SmartyFoot we used those license rights for SmartyPedo. We could have made SmartyPedo fully independent from SmartyFoot.
<u>[13] QUBCO Attorney:</u> Did you have any intention to deceive the USPTO?
<u>[14] Quinn:</u> Not at all. The joint venture agreement had nothing to do with SmartyPedo.
<u>[15] QUBCO Attorney:</u> Did you have any intention to deceive JUGCO?
<u>[16] Quinn:</u> No. JUGCO is an honorable company. We respect their highly innovative products. We are very impressed by their innovation process. However we are in different businesses and our hopes for joint ventures did not work out. No blame for them. No blame for us. Just was not a good marriage.
<u>[17] QUBCO Attorney:</u> Thank you. No further questions.

Deposition from the plaintiff's witness
Questioned by defendant's council
(Story for cross examination)

[1] JUGCO Attorney: When JUGCO sued QUBCO for infringing SmartyFoot, it was your idea to settle. Correct? Or not correct?
[2] Quinn: Correct.
[3] JUGCO Attorney: It was also your idea to enter into a joint venture agreement with JUGCO. Correct? Or not correct?
[4] Quinn: Correct.
[5] JUGCO Attorney: You knew at that time that JUGCO was in the clothing business only. Correct? Or not correct?
[6] Quinn: Correct.
[7] JUGCO Attorney: You had no reason to expect that JUGCO could contribute to your product line. Correct? Or not correct?
[8] Quinn: We had hope.
[9] JUGCO Attorney: Correct? Or not correct?
[10] Quinn: Not correct.
[11] JUGCO Attorney: You used the joint venture agreement as bait to get license to SmartyFoot. Correct? Or not correct?
[12] Quinn: That is a very offensive question.
[13] JUGCO Attorney: Correct? Or not correct?
[14] Quinn: Not correct.
[15] JUGCO Attorney: Thank you. No further questions.

Deposition from the defendant's witness
Questioned by defendant's council
(Story for direct testimony)

[1] QUBCO Attorney: Please tell us your name and spell it for the record.
[2] Rowan: Rowan R O W A N.
[3] QUBCO Attorney: What is your occupation?
[4] Rowan: I'm the president of JUGCO.
[5] QUBCO Attorney: How did JUGCO get involved with QUBCO?
[6] Rowan: QUBCO's SmartFoot product was a very good idea but did not work very well and cost too much because they did not know enough about clothing manufacturing. So, we made crucial improvements to make SmartyFoot better and less expensive. QUBCO did not patent SmartFoot so we patented SmartyFoot.
[7] QUBCO Attorney: Then QUBCO copied SmartyFoot and JUGCO sued QUBCO for infringing the SmartyFoot patent. What happened then?
[8] Rowan: If we didn't sue, then some huge greedy company might take advantage of us later. So we did sue. QUBCO offered to settle. We thought that was fair because after all SmartFoot was their idea.
[9] QUBCO Attorney: How did the joint venture come about?
[10] Rowan: At the same time QUBCO offered to settle they also offered a joint venture agreement. We thought that was a good idea because we thought there might be opportunities for combining smart products with clothing products.

[11] QUBCO Attorney: Did you think that the joint venture was bait to get license rights to SmartyFoot?
[12] Rowan: At the time I did not think that. As things turned out I did, and do, think that the joint venture was just bait and was not intended to work.
[13] QUBCO Attorney: Why did JUGCO copy SmartyPedo?
[14] Rowan: JUGCO believes that the SmartyPedo patent is not valid because QUBCO did not inform the USPTO about the joint venture.
[15] QUBCO Attorney: How can that be since the arbitrator voided the joint venture contract?
[16] Rowan: We have great experience with arbitrators and know that arbitrators can rule on the contract but can not rule on procedures in the USPTO. The USPTO had a right to know about the joint venture.
[17] QUBCO Attorney: Why would this be important to the USPTO?
[18] Rowan: In our experience with the USPTO we found them to be very fussy about who all is included as inventors, about everyone having a hand in the invention.
[19] QUBCO Attorney: Why do you think that QUBCO intended to deceive the USPTO?
[20] Rowan: JUGCO has been in business for three generations now and we have great experience with many other businesses, suppliers, customers, all sorts. We know how to recognize patterns of deceit. QUBCO's behavior is a classic case. They used the joint venture as bait to get license rights to SmartyFoot and then did not

tell the USPTO about the agreement to get the SmartyPedo patent. Deceit upon deceit.
<u>[21] QUBCO Attorney:</u> Thank you. No further questions.

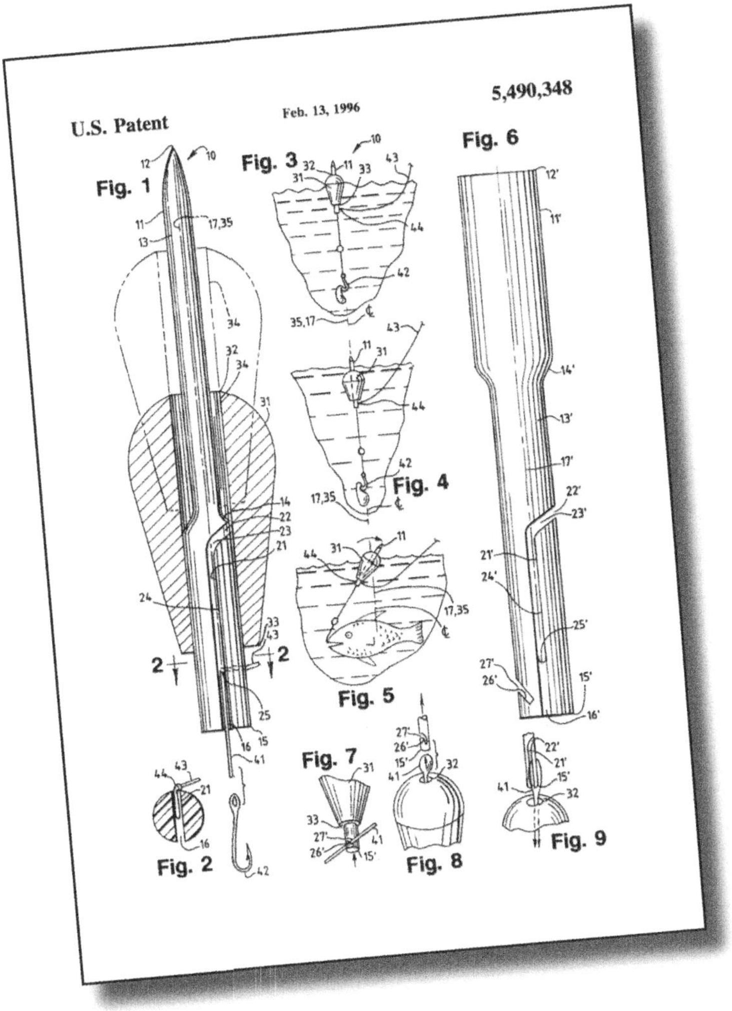

Deposition from the defendant's witness
Questioned by plaintiff's council
(Story for cross examination)

[1] QUBCO Attorney: The arbitrator ruled that the joint venture did not cover SmartyPedo. Correct? Or not Correct?
[2] Rowan: Correct.
[3] QUBCO Attorney: So, the joint venture is not relevant to the SmartyPedo patent. Correct? Or not Correct?
[4] Rowan: That's for the patent office to decide.
[5] QUBCO Attorney: Correct? Or not Correct?
[6] Rowan: I don't know.
[7] QUBCO Attorney: You are not a mind reader. Correct? Or not Correct?
[8] Rowan: Not correct.
[9] QUBCO Attorney: You claim that you can read minds?
[10] Rowan: I know how to recognize deceit.
[11] QUBCO Attorney: But, you don't know if QUBCO intended to deceive the USPTO. Correct? Or not Correct?
[12] Rowan: Not correct.
[13] QUBCO Attorney: Thank you. No further questions.

Can you identify some innovation wrongs which led to this trial?

Can you suggest some ways those innovation wrongs could have been put right?

If you were head of QUBCO what might you do to improve QUBCO's innovation rights?

If you were head of JUGCO, what might you do to improve JUGCO's innovation rights?

Appendices

Appendix A: Patent Example

US006112331A

United States Patent [19]

Horn

[11] **Patent Number:** **6,112,331**

[45] **Date of Patent:** **Sep. 5, 2000**

[54] **LIFTING FINGERS GUARD**

[76] Inventor: **Vernon W. Horn**, 9825 S. Charles, Chicago, Ill. 60643

[21] Appl. No.: **09/368,091**

[22] Filed: **Aug. 3, 1999**

[51] **Int. Cl.**[7] **A41D 19/00**

[52] **U.S. Cl.** **2/163**; 2/21; 294/25

[58] **Field of Search** 2/20, 21, 17, 160, 2/161.5, 161.6, 161.8, 163, 910; 294/25, 26; 473/59, 60, 61; 132/2, 213; 401/7; 602/22

[56] **References Cited**

U.S. PATENT DOCUMENTS

1,887,278	11/1932	Auster .	
2,769,179	11/1956	Love	2/21
2,881,022	4/1959	Brust	2/21
3,038,723	6/1962	Bergendorf	2/21
3,344,436	10/1967	Stubbs .	
3,398,951	8/1968	Disko	2/21
3,606,614	9/1971	Dimitroff .	
4,531,241	7/1985	Berger .	
4,651,350	3/1987	Dawiedczyk .	
5,131,095	7/1992	D'Amato	2/21
5,479,660	1/1996	Najac .	
5,487,188	1/1996	Micheloni .	
5,803,821	8/1998	Lupinacci	473/205

Primary Examiner—John J. Calvert
Assistant Examiner—Katherine Moran
Attorney, Agent, or Firm—Don Moyer

[57] **ABSTRACT**

The guard positions plates at the palm side of the two middle fingers and at a portion of the palm adjacent to the middle fingers thus distributing the weight of an object being lifted over the area of the plates in order to reduce pressure which would otherwise be present on smaller areas of the fingers and palm, without interfering with the remaining fingers which provide control of the lifted object.

4 Claims, 1 Drawing Sheet

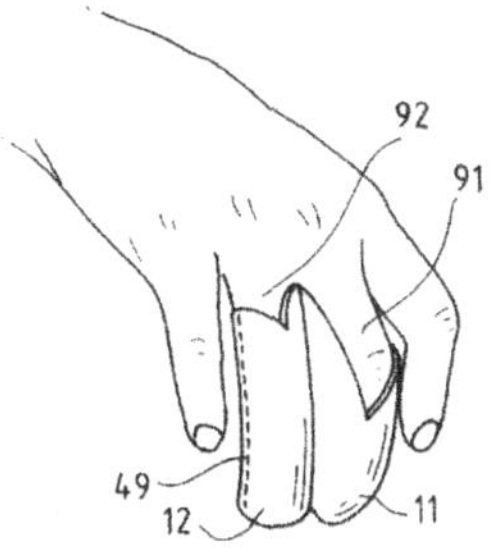

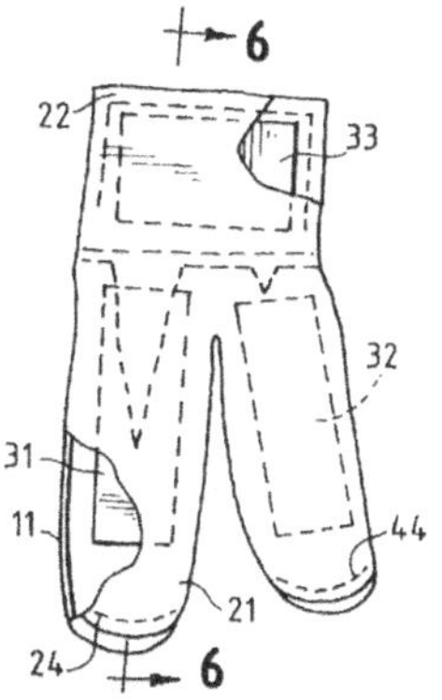

Published Patent Face Page

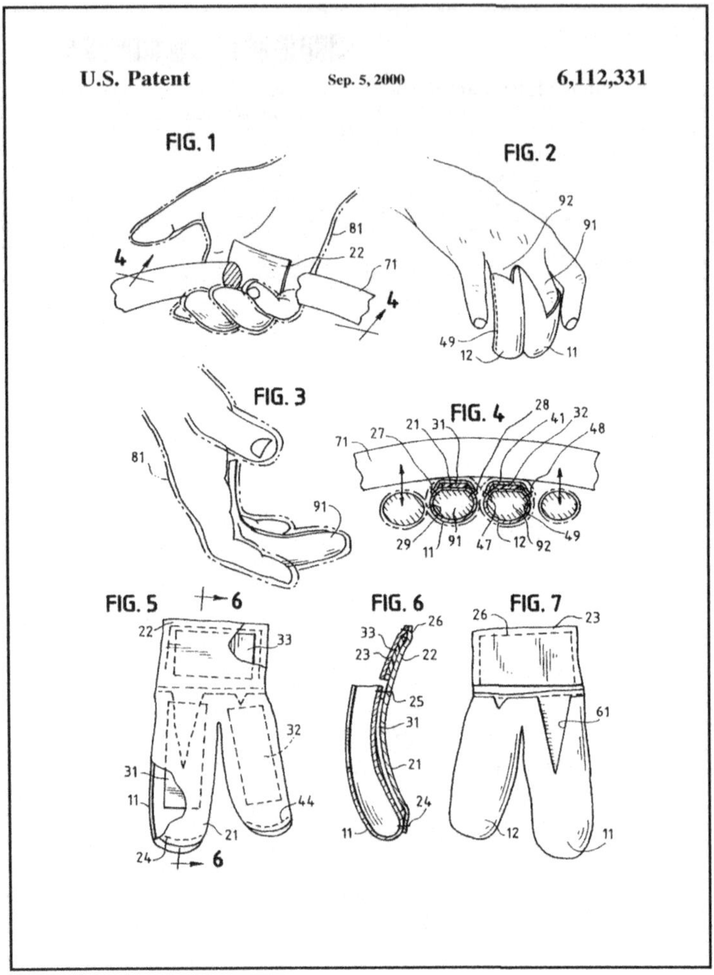

Drawings Sheet

6,112,331

1

LIFTING FINGERS GUARD

BACKGROUND OF THE INVENTION

The lifting fingers guard positions plates along the palm side of the two middle fingers and along the portion of the palm adjacent to the two middle fingers, while leaving remaining fingers free, so that the plates distribute the weight of objects being lifted over the area of the plates thus reducing the pressure which otherwise would make lifting difficult.

The invention is based on two discoveries. The first discovery is that plates can be positioned at the two middle fingers and a portion of the palm adjacent to the two middle fingers so that the weight of objects being lifted is distributed over the area of plates making it easier to lift objects because the pressure on smaller areas of the fingers and palm is greatly reduced. The second discovery is that plates should be only positioned on the two middle fingers and not the other fingers because the mobility of fingers other than the middle two fingers provide control of the object being lifted and must not be interfered with.

Various guards worn on all or part of a hand are shown in prior art; for example, by Auster in U.S. Pat. No. 1,887,278, by Stubbs in U.S. Pat. No. 3,344,436, by Dimitroff in U.S. Pat. No. 3,606,614, by Berger in U.S. Pat. No. 4,531,241, by Dawiedczyk in U.S. Pat. No. 4,651,350, by Najac in U.S. Pat. No. 5,479,660, and by Micheloni in U.S. Pat. No. 5,487,188. None of these—and no combination of these—suggests plates positioned just under the two middle fingers and on a portion of the palm adjacent to the two middle fingers which would distribute the weight of an object being lifted over the area of the plates.

SUMMARY OF THE INVENTION

One form of this lifting fingers guard comprises a ring finger plate, the ring finger plate being positioned at a ring finger palm side; a palm plate connected to the ring finger plate; and a middle finger plate connected to the palm plate, the middle finger plate being positioned at a middle finger palm side, the middle finger plate, the palm plate, and the ring finger plate not interfering with motions of remaining fingers.

Alternate forms and objects of the invention will be comprehended in the drawings and description, which will make other alternate forms and objects obvious hereafter to persons skilled in the art.

BRIEF DESCRIPTION OF THE DRAWINGS

FIG. **1** shows an object being lifted with the lifting fingers guard in place.

FIG. **2** shows the guard seen from above a hand.

FIG. **3** shows the guard seen from beside a hand.

FIG. **4** shows a cross section viewed along line **4—4** in FIG. **1**.

FIG. **5** shows the lifting side of the guard.

FIG. **6** shows a cross section viewed along line **6—6** in FIG. **5**.

FIG. **7** shows the non-lifting side of the guard.

DESCRIPTION OF THE PREFERRED EMBODIMENTS

One form of the lifting fingers guard comprises a middle finger sleeve **11** and a ring finger sleeve **12** into which a person's middle finger **91** and ring finger **92** respectively can

2

be placed. The sleeves are made of a durable material such as leather. On the lifting side—that is, the palm side—the sleeves have pocket forming layers, **21** and **41** respectively which form a middle pocket and a ring pocket respectively, also of a durable material. A middle finger plate **31** and a ring finger plate **32**, are inserted respectively in the middle pocket and the ring pocket formed by the pocket forming layers.

The middle finger sleeve **11** and the ring finger sleeve **12** are connected to a palm pad **22** which has a palm pocket forming layer **23** forming a palm pocket into which a palm plate **33** is inserted.

The three plates—**31**, **32**, and **33**—though bendable, are hard so that they distribute the weight of an object being lifted, such as the rod **71** shown in FIG. **1** and FIG. **4**, over the area of the plates. This reduces the pressure which would otherwise be present on smaller areas of the middle finger, the ring finger, and the palm portion.

Stitching is indicated at **24**, **25**, **26**, **27**, **28**, **29**, **44**, **47**, **48**, and **49** to show how the sleeves, the pocket forming layers, and the palm pocket can be put together. Other attaching means, such as adhesives, can also be used in place of stitching. The sleeves and the palm pad with the pockets could be made of various plastic materials with the elements heat welded together. Also, the plates could be heat welded onto the sleeves and the palm pad, and could be integral parts of the palm sides of the sleeves and of the palm pad.

The sleeves are shown covering the ends of the fingers, which is preferred when rough objects are lifted. In cases where the objects being lifted are not rough, the finger ends can be left off. The guard can be worn under a light glove, which does not interfere with motions of the other fingers, as indicated by **81**.

In FIG. **7** a deep notch **61** is shown cut into the non-lifting side of the middle finger sleeve. This allows for more free movement of the middle finger knuckle as shown in FIG. **2**. Since finger sizes vary this notch, and possibly a notch on the ring finger sleeve, can be custom made.

Many alternate forms of the lifting fingers guard are possible which are consistent with the invention and with the two discoveries which are the basis of the invention. Thus alternate forms must have a middle finger plate, a ring finger plate, and a palm plate positioned on the lifting sides—that is palm sides—of the middle finger and the ring finger and a portion of the palm adjacent to these fingers and must not interfere with the remaining fingers.

The function of the sleeves and of the palm pad is to position the middle finger plate, the ring finger plate, and the palm plate on at the palm side of the fingers and palm portion. A middle finger plate connected to a palm plate and a ring finger plate connected to the palm plate could be positioned at the lifting sides of the middle fingers by various alternate mechanisms.

For example, these connected plates could be positioned by use of adhesive. They could be worn under a light glove which does not interfere with the other fingers and thumb. Alternatively a ring which at least partially encircles a middle finger and which is attached to the middle plate can position the plates. Similarly, a ring which at least partially encircles a ring finger and is attached to the ring finger plate can position the plates. And, a first ring and a second ring at least partially enciriling respectively a middle finger and a ring finger and attached respectively to the middle finger plate and to the ring finger plate can position the plates.

Alternate forms for the plates, and other ways to position the plates will be obvious hereafter to persons skilled in the art. Therefore this invention is not limited to the particular examples shown and described here.

Written Description

6,112,331

3

I claim:

1. A lifting fingers guard comprising:

a ring finger plate, the ring finger plate being positioned at a ring finger palm side when in use;

a palm plate connected to the ring finger plate; and

a middle finger plate connected to the palm plate, the middle finger plate being positioned at a middle finger palm side, the middle finger plate, the palm plate, and the ring finger plate being hard to distribute held weight, the middle finger plate, the palm plate, and the ring finger plate not interfering with motions of remaining fingers when in use.

2. The device of claim 1 wherein the ring finger plate is combined with a ring finger sleeve, the middle finger plate is combined with a middle finger sleeve, the palm plate is combined with a palm pad, with the palm pad being connected to the ring finger sleeve and to the middle finger sleeve.

3. The device of claim 2 wherein the ring finger plate is combined with the ring finger sleeve by being inserted into a ring pocket on the ring finger sleeve, the middle finger

4

plate is combined with the middle finger sleeve by being inserted into a middle pocket on the middle finger sleeve, and the palm plate is inserted into a palm pocket on the palm pad.

4. A lifting fingers guard comprising:

a ring finger sleeve positioned on a ring finger when in use;

a ring pocket on the ring finger sleeve;

a ring finger plate inserted into the ring finger pocket;

a palm pad connected to the ring finger sleeve;

a palm pocket in the palm pad;

a palm plate inserted into the palm pocket;

a middle finger sleeve connected to the palm pad and positioned on a middle finger when in use;

a middle finger pocket on the middle finger sleeve; and

a middle finger plate inserted into the middle finger pocket.

* * * * *

Claims: The Claimed Invention

Appendix B: Patent Face Page

The face page of a published patent (see Appendix A) contains much useful information. First, notice the numbers in brackets. These are international identifier (INID) codes which tell what type of information follows the INID code.

For example, INID code 76 indicates the inventor, INID code 22 indicates the date the patent application was filed, and INID code 45 indicates the date the patent was issued. Those dates are just over a year apart. Nowadays the time between filing and issuing is usually longer than three years.

INID code 51 indicates the international classification for the patent and INID 52 indicates the US classifications for the patent, the bold classification being the primary classification and the others being classifications where the patent is cross-referenced. The USPTO is working with other patent offices to devise one classification system using the best practices of all.

INID code 58 indicates the classifications where the examiner searched thinking that similar inventions in those classifications might anticipate Vernon's invention.

INID code 56 indicates patents, and sometimes non-patent documents, which disclose things like Vernon's invention. Some of these references were supplied by the inventor and some by the examiner. Nowadays the references cited by the examiner are indicated with a star * mark. Appendix D has more information about the classification system and about searching.

The abstract was supplied as part of the patent application. The patent face has a representative drawing. In this case two drawings are there.

Remember in the General Introduction the remaining parts of the published patent were explained:

1) After the face page is a drawing sheet, sometimes several sheets.

2) After the face page is the written description.

3) The claims are last.

In the example the written description and the claims are each on separate pages, which is unusual but nice for an example.

Appendix C: Steps to a Patent

It starts with an idea which can provide a new result, to improve health and safety, to make life and work easier, to start and grow a business. But: Will it work? Will anyone buy it? Is it new?

Good idea to make a model to see if it will work. Sooner or later you will need to either have a working model or drawings which can be used to make a working model, because to get a patent you need to describe the thing so that a person with appropriate skill can make and use the thing. Remember our enable trials.

How could you find out if anyone will buy? Don't tell them your idea and ask, they might be too nice to tell you it's a lousy idea. And, telling them could lead to losing any patent right.

Your idea achieves a new result, so you could ask people what features the new result should have for them to buy and use. If your idea has all those features you might have a winner. If your idea does not have all those features go back to the drawing board.

It's best to get a professional patent search done before applying for a patent. A professional search is a small part of the whole cost of a patent so not doing this is false economy. It's best to get the search done via a patent agent or patent attorney.

(What's the difference between patent agents and patent attorneys? For one thing there are no known jokes about patent agents. Both are licensed by the USPTO after passing a beastly exam. Both can do the same work with the USPTO. A patent attorney is also a lawyer and can do other lawyer work, such as sue you when you don't pay.)

Back to the topic. Patent agents and patent attorneys know professional searchers and understand the quality of the search. Persons trying to do patent work without getting good professional help can create a terrible mess which is likely to lose any patent right.

You can and should do some searching yourself. You might quickly find that your idea is not novel and save lots of money. Appendix D has information about searching.

Results of your and a professional search can be very informative and can give you ideas for improving your idea. You might find that your idea is not new, but that the already patented inventions do not provide the new result you want your idea to provide. This is an opportunity to go back to the drawing board and improve your idea until it succeeds where an already patented invention did not succeed.

It works. It will sell. It's new. So, it's time to file a patent application. You can do this yourself. You can also do brain surgery on yourself. In both cases you are likely to make a mess. Get good professional help! There are several ways to tell if a person is likely to be good help:

1) Does that person listen carefully and ask hard probing questions to be sure to understand your idea?

2) Does that person tell you all the steps and get you to participate at each step?

Your application for a patent is filed. Now you wait maybe two years until a patent examiner takes up your application. Appendix D will describe some of what the examiner does. Then you will get a first office action by your examiner. The examiner is likely to object to and reject parts of your application. This should lead you and your patent agent or patent attorney to make changes and explanations which will improve your patent.

Those changes and explanations are filed and you wait again. If everyone has done a good job, and nothing unexpected was found by the examiner, you will get a notice of allowance. You pay the issue fee and soon you have a patent granting you the right to exclude as was discussed in the General Introduction.

Appendix D: USPTO Patent Classification

When a patent examiner takes up your patent application a key step is searching the prior art to see if your claims are anticipated. In patent lingo "prior art" is any information available to the public. The biggest and best organized body of technology information is US published patents and published patent applications. So, the patent examiner will search this.

Because you wanted to anticipate what the examiner would find you also had a professional search of this done before your patent application was written. And, you did some of your own searching.

Here are some examples of using the USPTO patent classification for prior art searching.

Go to the USPTO web site: USPTO.gov. When you're there look for the site index, which now is in the upper right portion. Click on than and then click on "i" and find the "Index, Patent Classification" entry and go there. There are several menu bars across the top area. Find "Class Numbers and Titles" and go there.

Web sites change so these directions might be hard to follow on a revised version of the USPTO web site. You will still be able to find a path to the class numbers and titles.

The class numbers and titles page shows the list of main classes. Class "002" is for apparel. Go there and you'll see the many sub-classes of apparel. Scroll down to find the "HAND OR ARM COVERINGS" heading. Major categories in a class are in all caps. A sub-class preceded by a single dot depends directly from the main category. A sub-class preceded by two dots depends from the sub-class preceded by one dot above it. Etc.

Scroll down to find the "Finger" sub class which would be written 2/163 and which is the main class for Vernon Horn's patent which is shown in Appendix A. To the left is a small blue "A" which is a link to published patent applications and a small red "P" which is a link to issued patents. Take the red "P" link and you see all the patents in that class/ sub-class. Vernon's (6,112,331) is now way down the list. Click on any of the patents listed and you see the patent.

Here's another example.

Suppose you "invented" a cap rack which keeps the cap brim furled and this is your drawing in your notebook:

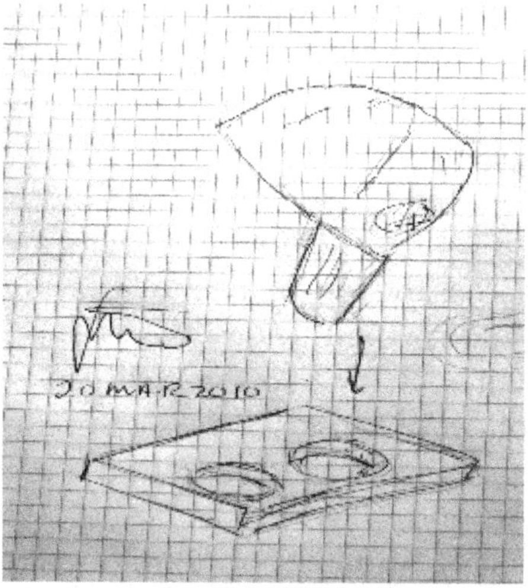

Again go to the USPTO web site: USPTO.gov. When you're there look for the site index, which now is in the upper right portion. Click on than and then click on "i" and find the "Index, Patent Classification"

entry and go there. There are several menu bars across the top area. This time find "USPC Index" and go there. This is the index to the classification, which is an alphabetical index of names of things.

The names are archaic (the patent system started with Thomas Jefferson remember) so you must be a good detective to find things. T-shirts are under "nether garments" for example.

Your invention is a cap rack so lets start looking under "R" to see if there is a rack for hats. There it is 211/30. Click on the "30" link to go there. Scroll down the titles to see if there is anything relevant. Better look at 5,758,779 a "Brim forming cap holder" so click on that. Your browser might not have the plug-in to see the USPTO images so you can use Google patent search to find the patent. Enter the patent number and see this:

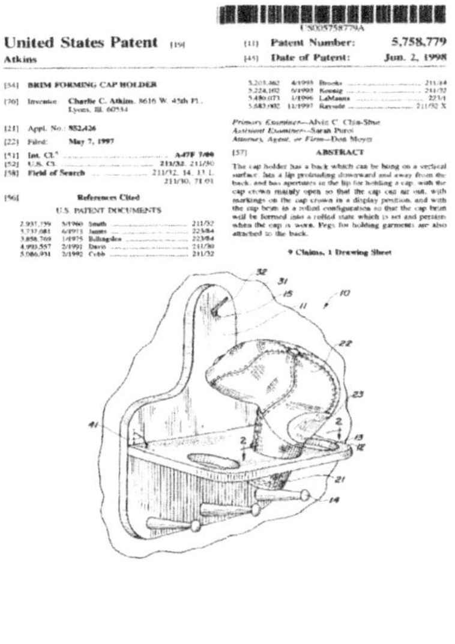
United States Patent [19]
Atkins

[11] Patent Number: 5,758,779
[45] Date of Patent: Jun. 2, 1998

[54] BRIM FORMING CAP HOLDER

[76] Inventor: Charlie C. Atkins, 8616 W. 45th Pl., Lyons, Ill. 60534

[21] Appl. No.: 852,426

[22] Filed: May 7, 1997

[51] Int. Cl.⁶ A47F 7/00
[52] U.S. Cl. 211/32; 211/30
[58] Field of Search 211/32, 14, 13.1, 211/30, 71.01

[56] References Cited

U.S. PATENT DOCUMENTS

2,937,759	5/1960	Smith	211/32
3,737,081	6/1973	James	223/84
3,858,769	1/1975	Billingsley	223/84
4,993,557	2/1991	Davis	211/30
5,086,931	2/1992	Cobb	211/32
5,203,462	4/1993	Brooks	211/14
5,224,162	6/1993	Koenig	211/32
5,480,073	1/1996	LaManna	223/1
5,683,002	11/1997	Raysule	211/32 X

Primary Examiner—Alvin C. Chin-Shue
Assistant Examiner—Sarah Purol
Attorney, Agent, or Firm—Don Moyer

[57] ABSTRACT

The cap holder has a back which can be hung on a vertical surface, has a lip protruding downward and away from the back, and has apertures in the lip for holding a cap, with the cap crown mainly open so that the cap can air out, with markings on the cap crown in a display position, and with the cap brim in a rolled configuration so that the cap brim will be formed into a rolled state which is set and persists when the cap is worn. Pegs for holding garments are also attached to the back.

9 Claims, 1 Drawing Sheet

OK. Back to the drawing board!

Appendix E: Acknowledgements

Thanks to leadership at John Marshall Law School for making possible the class there for the trailblazing law student mock trial coaches.

Thanks to the John Marshall Law School student trailblazers who coached the trailblazing mock trials.

Thanks to leadership at Austin Polytechnical Academy for welcoming the trailblazing mock trials there.

Thanks to the Austin Poly students who participate in the trailblazing mock trials.

Thanks to all at the Center for Labor and Community Research who host the trailblazing mock trials at APA and provide all sorts of support.

Patent drawings scattered throughout here are the work of the firm Shutters and Stade. See: www.penciltoink.com

About the Author

Donald F. Moyer, PhD

Physics teaching and research and other higher education must-do stuff was Moyer's first career. This aroused interest in mysteries of innovation which led to the second career of History of Science teaching and research and related pursuits.

Interest in innovation then led to a third career. Accepting challenges from the Energy Related Inventions Program of the US Departments of Commerce and Energy and from the Harold Washington administration in Chicago he started a not-for-profit to aid independent inventors in 1983.

As funding a not-for-profit became increasingly difficult he passed the exam to be licensed by the USPTO as a patent agent and was the agent for forty issued patents from 1996 until alleged retirement in 2004.

Since 1998 he has been responsible for the Patent Clinic at John Marshall Law School.

Innovation Law Mentoring – created in 2009 - is the origin of the scenarios here.

Donald F. Moyer, PhD

- is an unreconstructed physicist,

- is an Historian of Science and Technology,

- is a Patent Agent licensed by the U. S. Patent and Trademark Office since 1994 specializing in health, safety and assistive technologies,

- is an adjunct professor at John Marshall Law School,

- received his Ph.D. from the University of Wisconsin,

- taught and did research at various institutions including University of Wisconsin, IIT, Northwestern, Fermilab, and Texas A & M,

- published extensively on topics in Physics, History, Philosophy, Sociology, and Alzheimer's,

- was a frequent consultant, invited speaker, and organizer of Conferences, Colloquia, and Seminars on these and other topics,

- was elected to Sigma Xi and Sigma Pi Sigma,

- is a member of the American Physical Society,

- had research support including NDEA, NSF, MacArthur foundation, Welch foundation, and Ford Foundation grants,

- was still working as a physicist doing research with a device intended to improve radiation therapy before semi-retirement in 2004,

- was the agent for forty issued patents from 1996 to 2004,

- created Chicago's Inventors' Council funded by the city of Chicago and the US Department of Energy to work for for many years with independent inventors and small businesses to foster innovation diffusion,

- has been teaching patent writing at John Marshall Law School since 1998,

- created the John Marshall Law School Innovation Law Mentoring class,

- served on the City of Chicago Community Development Block Grant Executive Committee and chairing the Economic Development Sub-Committee thereof and on the boards of several not-for-profit organizations.

Made in United States
North Haven, CT
05 November 2026

21790617R00143